OFFSHORE BLOWOUTS

Causes and Control

Gulf Publishing Company
Houston, Texas

OFFSHORE BLOWOUTS

Causes and Control

PER HOLAND

OFFSHORE BLOWOUTS:
Causes and Control

Data for Risk Analysis in Offshore Operations Based on the SINTEF Offshore Blowout Database

Gulf Publishing Company
Book Division
P.O. Box 2608 ☐ Houston, TX 77252-2608

10 9 8 7 6 5 4 3 2

Library of Congress Cataloging-in-Publication Data

Holand, Per.
Offshore blowouts : causes and control / by Per Holand.
p. cm.
Includes bibliographical references and index.
ISBN 0-88415-514-5 (alk. paper)
1. Oil wells—Blowouts. 2. Gas wells—Blowouts. 3. Underwater drilling. I. Title.
TN871.215.H65 1997
622′.3382—dc21 97-20910
CIP

Contents

Preface

The book is based on the SINTEF Offshore Blowout Database which contains valuable information to be used for analyzing risk related to offshore installations. The database focuses on blowout causes and parameters important when developing risk analyses, particularly flow path, release points, flow mediums, ignition time, duration, and fatalities. Data from 380 blowouts are included.

The presentation of the blowouts in this book is, however, limited to the 124 blowouts that occurred on the Outer Continental Shelf of the U.S. Gulf of Mexico and the Norwegian and UK waters in the period from January 1980 till January 1994.

The blowout frequencies and associated trends for the various phases of operation have been investigated. The contribution from blowouts to the fatal accident rate is analyzed. Pollution caused by blowouts is discussed. Important factors when evaluating offshore risk, such as ignition probability, time to ignition, ignition trends, blowout duration, and blowout flow path, are also included.

Per Holand

Acknowledgment

I would like to acknowledge Professor Marvin Rausand at the Norwegian University of Science and Technology (NTNU) first for his assistance with this book and the Ph.D. thesis the book is based upon, and secondly for the work he has carried out with respect to safety and reliability research and education within SINTEF and at the NTNU.

Further, I would like to acknowledge NTNU for allowing me to publish a book based on my Ph.D. thesis.

I also acknowledge the sponsors of the SINTEF Offshore Blowout Database, because they let me use the database as the basis for the thesis. Further, I thank them because their use of the database has been a very important quality control of the database.

Finally, I would like to thank my colleagues at SINTEF Safety and Reliability, and especially the secretaries who do not know the word no.

List of Abbreviations

2D	Two dimensional
3D	Three dimensional
API	Petroleum Institute (USA/International)
BOP	Blowout Preventer
CARA	Computer Aided Reliability Analysis
CCA	Cause Consequence Analysis
DEV.	Development
DHSV	Down Hole Safety Valve≈SCSSV
DNV	Det Norske Veritas
DRLG.	Drilling
ERCB	Energy Resources Conservation Board (Canada)
EXPL.	Exploration
FAR	Fatal Accident Rate
FTA	Fault Tree Analysis
GoM	Gulf of Mexico
HAZOP	Hazard and Operability
HP/HT	High pressure/High temperature
HSE	Health and Safety Executive (UK)
ILCI	International Loss Control Institute
ISRS	International Safety Rating System
MMS	Minerals Management Service (USA)
MORT	Management Oversight Risk Tree
MWD	Measurement While Drilling
NAF	Neal Adams Firefighters (USA)
NWTB	No. of wells to blowouts
NPD	Norwegian Petroleum Directorate
NTH	Norwegian Institute of Technology (before Jan. 1, 1996)
NTNU	Norwegian University of Science and Technology (from Jan. 1, 1996)

OARU	Occupational Accident Research Unit
OCS	Outer Continental Shelf
OSD	Offshore Safety Division (UK)
QRA	Quantified risk analysis
ROCOF	Rate of Occurrences of Failures
SCSSV	Surface Controlled SubSurface Safety Valve ≈DHSV
SEMP	Safety and Environmental Management Program (USA)
SMORT	Safety Management and Organization Review Technique
STEP	Sequentially Time Event Plotting
TLP	Tension Leg Platform
WOAD	World Offshore Accident Databank (Norway)
WR	Wireline retrievable

CHAPTER ONE

Introduction

WHY RISK ANALYSIS?

Exploration and development of offshore oil and gas fields involve a number of risks related to loss of human lives, pollution, and loss of material assets. All those involved in the offshore industry are aware of the hazards. The potential for major accidents will always be present, but it is important to keep the risks within acceptable levels, and as low as reasonably practicable.

A main contributor to the total risk is uncontrolled release of pressurized hydrocarbons, i.e., gas leakages and blowouts. It should, however, not be forgotten that other aspects such as vessel stability, helicopter transport, and occupational accidents are also significant contributors to the total risk.

History shows that uncontrolled releases of hydrocarbons have caused several major accidents. The Bravo blowout on the Ekofisk field in 1977, the West Vanguard blowout in 1985, the Piper Alpha gas leak in 1988, and the Ocean Odyssey blowout in 1988 are all well-known accidents that occurred in the North Sea. In addition, several less severe accidents involving uncontrolled releases have occurred in the North Sea.

RISK ANALYSIS REGULATIONS

Because of the risks involved and the potential for major accidents, a variety of risk analyses of the offshore activities is now mandatory both in Norway and the United Kingdom (UK).

The Norwegian Petroleum Directorate (NPD) has issued a specific regulation [3] related to implementation and use of risk analyses in the petroleum activities. The purpose of the regulation is, through risk analyses, to establish and maintain a fully *satisfactory level of safety* for people, the environment, and assets and financial interests in the petroleum activities.

The *satisfactory level of safety* is described through *acceptance criteria.* These criteria are used to express an acceptable level of risk in the activities. The operator has to define the risk acceptance criteria before any risk analysis is carried out.

It is further stated that risk analyses shall be planned, carried out, used, and updated in a controlled manner. Attempts should be made to eliminate or reduce the individual risks identified through risk analyses. Probability-reducing measures shall, to the extent possible, be given priority over consequence-reducing measures.

In the UK, operators are now required to submit a so-called *Safety Case* [6] for each mobile and fixed installation. The Safety Case shall be submitted to the Health & Safety Executive (HSE) and is to be seen as a "living document" that is updated on a regular basis to take account of changing activities, technologies or other circumstances. The Safety Case regulations were based on the recommendations in the Lord Cullen Report [70], which were made after the Piper Alpha disaster in 1988. The regulations are similar to the Safety Case requirements, which are included in the latest revision of the so-called *Seveso Directive* [19].

The Safety Case regulations state that all hazards with the potential to cause a major accident shall be identified, their risks evaluated, and measures taken to reduce the risks to persons to as low as reasonably practicable (the so-called ALARP principle) [6]. Incidents causing major accidents are, among others, fire, explosion, the release of dangerous substances involving death or serious personal injury, or other events causing major damage to the installation. It is further stated in the Safety

Case regulations that the likelihood of the major hazards should be assessed.

In the U.S., there are so far no regulations concerning risk analyses of the offshore activities [59]. However, prompted by the Piper Alpha disaster, the Minerals Management Service (MMS) conducted a series of reviews of offshore safety practices and of the MMS regulatory program. MMS is the U.S. Outer Continental Shelf (OCS) regulatory authority. A Marine Board study concluded that:

- OCS operators should improve the "human and organizational factors" in their safety programs, and
- MMS should refocus its regulatory and inspection programs to emphasize the critical role of human, organizational, and management influences on safety and environmental protection.

This study prompted the formation of the Safety & Environmental Management Program (SEMP). The purpose of the program is to reduce the risk of accidents and pollution from U.S. OCS operations.

The industry asked MMS to postpone the new regulations and instead let them try to develop a voluntary approach. In response to MMS's SEMP proposal, the American Petroleum Institute (API) developed a new recommended practice, API RP75, "Recommended Practices for Development of a Safety and Environmental Management Program for Outer Continental Shelf (OCS) Operations and Facilities" [9]. API RP75 was published in May 1993. MMS will determine whether to continue the voluntary program or to consider other alternatives in mid-1997.

API also developed a companion document, API RP14J [7], which outlines the hazards and risk analyses to be carried out as part of SEMP.

IMPORTANCE OF HIGH QUALITY RISK ANALYSIS

When performing a *quantified risk analysis* (QRA), there are three important aspects that are required to ensure a satisfactory quality of the analysis:

- A thorough understanding of the system and its operational characteristics
- The proper use of risk analysis methods
- Risk input data of high quality

A risk analysis is usually compiled by using different models and risk input data. Both the models and the data will be more or less incomplete. It is important that the analysis approximately reflects the real risk picture, and, further, that it is possible to identify and rank the major risk contributors. If not, the analysis has limited or no value.

Risk analyses of offshore installations are carried out on various levels. In the concept phase, the concept is analyzed. The results from the analyses have impact on the final concept layout. As the project proceeds, risk analyses are performed at different levels for various applications. A risk analysis is to be seen as a "living document" that is updated on a regular basis to take account of changing activities, technologies, or other circumstances.

Risk analyses may have large effects on the system layout and may indicate whether or not companies should take protective measures. For example, there are two relatively new Tension Leg Platforms (TLPs) in the Norwegian sector. These installations are, in principle, similar. To determine whether fire insulation of the risers was required to meet the risk acceptance criteria, consultants (one for each TLP) conducted risk analyses. The results of the analyses indicated that only one of the TLPs needed riser fire insulation. For the TLP on which fire insulation was not required, the total installation investment was reduced at least (US) $120 million [49]. This TLP was installed two years after the first one was installed. New and updated SCSSV (Surface Controlled Subsurface Safety Valves) reliability data were the main reason the consultants arrived at different conclusion (i.e., the updated reliability data showed that the SCSSV reliability had improved compared to previous experience).

A good question to ask is, could an acceptable risk level be achieved by other means than the costly fire insulation (i.e. reducing the blowout probability or reducing the probability that a blowout could lead to an oil spill and subsequent fire at sea)?

The previous example illustrates the importance of a proper risk analysis, and further, that good input data are important both with respect to estimating the risk itself and to assessing the effects of various risk-reducing measures.

This book concerns the blowout experience from the U.S. GoM OCS and the Norwegian and UK sectors of the North Sea, 1980-1994. The experience is stored in the SINTEF Offshore Blowout Database [61]. Information from the database is presented in the section *SINTEF Offshore Blowout Database* on page 33.

CHAPTER TWO

Fatal Accident Rates and Blowouts

INTRODUCTION

Most people assume that blowouts are assumed to be one of the *major contributors to risk* in offshore activities [58].

Risk in offshore activities is normally related to:

- Loss of human lives
- Pollution of the environment
- Loss of material assets

Regarding *loss of material assets*, blowouts seem to be a major contributor to the total risk. Out of the 118 blowouts (not including blowouts from external causes) that occurred in the U.S. GoM and the North Sea from 1980-1994, fourteen of the installations were categorized as total loss or severely damaged. Of these fourteen blowouts, twelve blowouts ignited while two did not. The fire itself was the main cause of the damages for these twelve incidents. The two blowouts that did not ignite caused a subsea crater, which caused one installation to sink and the other to tilt. Besides the listed costly consequences, blowouts always cause severe time losses, and often the well has to be plugged, abandoned, and redrilled. In general, blowouts are costly incidents (see Chapters 6-10 for more details).

In terms of *pollution of the environment*, none of the blowouts in the North Sea or the U.S. GoM OCS from 1980-1994 involved *large* releases of oil/condensate into the sea. The most severe incidents were reported with 10 m^3 (63 bbl) of oil to the sea, some few cubic meters of oil to sea and large sheens. Large release incidents caused by blowouts have, however, occurred during other periods and in other areas. Blowout incidents that resulted in large releases from 1970 till 1994 comprise:

- September 1992, U.S. (Not on the OCS)
- May 1989, Caspian Sea
- October 1987, Mexico
- May 1986, Venezuela
- October 1984, Indonesia
- March 1983, Venezuela
- February and March 1983, Iran, two blowouts as a result of the Iranian /Iraqi war
- October 1980, Saudi Arabia
- February 1980, Spain
- January 1980, Nigeria
- June 1979, Mexico
- February 1978, Iran
- April 1977, Norway
- August 1973, Trinidad
- December 1971, Iran

The blowout in Nigeria in January 1980 was the most serious incident of all. The oil polluted islands and channels of the Niger delta with 30,000 tons (220,000 bbl) of crude oil, ruining the food supplies for thousands of Nigerian fishing people. It was claimed by the Nigerian government that 180 people died due to pollution of the drinking water [53]. The operating company, Texaco, stated however that detailed studies found no evidence whatsoever of any fatalities directly resulting from the blowout or the oil spill.

Although the probability of experiencing large hydrocarbon releases as a result of a blowout seems low based on the experiences in the North Sea and the U.S. GoM OCS, experience from other areas shows that large hydrocarbon releases *may* also occur in the U.S. GoM OCS and the North Sea (see Chapters 6-10 for more details).

According to a 1993 MMS press release, statistics show that 45% of the oil that enters the world's seas is caused by tankers and other marine vessels; 53% is caused by municipal and industrial runoff; and only about 1.5% of the oil that enters these seas is due to offshore oil and gas production [60].

Bravo blowout, Norwegian sector of the North Sea, 1977
During a workover operation, the oil well started to flow. The crew did not control the initial oil flow, and a major oil spill was the result. During the eight days of the blowout more than 20,000 m^3 (125,000 bbl) had spilled.

PERSONNEL RISK INDICES

Personnel accident risk is usually estimated as the observed number of injuries or fatalities per time unit. In the blowout context, such estimates related to personnel injuries are of limited relevance. The number of injuries caused by blowouts is insignificant compared with injuries in general.

Fatal Accident Rate (FAR) is the most common North Sea measure regarding fatalities per time unit. FAR is the estimated number of fatalities per 10^8 exposure hours [24].

In industrial risk analyses, the number of working hours is used as the exposure time for calculating the FAR value. For offshore risk analyses, the FAR value may be calculated based on the *actual working hours* or the *total hours the personnel are on the installation.* FAR is also used for other applications, such as air traffic, in which the exposure time used is the passenger hours.

The FAR value is frequently used for defining the acceptance criteria when evaluating offshore risk. It is then important to note whether the

FAR value is based on the working hours or the total hours the personnel are on the installation.

The FAR value is a reasonable measure for risk analyses. If used properly, the FAR value will give a fair representation of the actual fatality risk. Nevertheless, it is important to note that estimated FAR values for specific installations (offshore platform or a chemical plant) can never be verified through experienced fatality statistics. This is because low probability incidents with a high number of fatalities will commonly have a high influence on the estimated FAR value. If one such incident occurs, the experienced FAR will be much higher than the estimated FAR and vice versa. This is illustrated in Table 2.3 where the effect of the Piper Alpha and Alexander Kielland accidents completely changed the FAR values for the entire North Sea.

Piper Alpha gas leak, UK sector of the North Sea, 1988
The explosion on the Piper Alpha production platform resulted in a fire that completely destroyed the platform and cost 167 lives and millions of dollars a day in lost revenue.

Alexander Kielland accident, Norwegian sector of the North Sea, 1980
During bad weather the Alexander Kielland semisubmersible rig, which was used as a living quarter on the Ekofisk field, capsized when a bracing broke off. Of the 212 men on board, 123 lost their lives.

FATALITIES IN OFFSHORE ACTIVITIES

Seven of the 118 blowouts in the North Sea and the U.S. GoM OCS from 1980-1994 resulted in fatalities (not including blowouts caused by external forces). Two North Sea blowouts resulted in a total of two fatalities, while five U.S. GoM OCS blowouts caused a total of 18 deaths.

Table 2.1 shows the total number of fatalities. Table 2.1 also includes fatalities associated with helicopter transport of personnel.

Table 2.1
Total Number of Fatalities in the North Sea and the U.S. GoM OCS January 1980-January 1994

Area	No. of fatalities 1980-1994 *
Norway	150**
UK	274***
US GoM OCS	144
Total	**568**

* based on [45], [52], [1], and [2]
** includes the Alexander Kielland accident with 123 fatalities
*** includes the Piper Alpha accident with 167 fatalities

Observe that the number of fatalities in Table 2.1 is different from the fatality statistics in the World Offshore Accident Databank (WOAD) [73], in which 487 offshore accident fatalities were reported during the same period. The sources of information used in Table 2.1 are regarded to be more accurate.

The SINTEF Offshore Blowout Database show that in total, 20 persons were killed in blowouts in the same areas and period (i.e., 3.5% of the total number of fatalities were caused by blowouts). Disregarding the Piper Alpha and Alexander Kielland accidents, the total would be 278 fatalities (i.e., 7.2% of the deaths were caused by blowouts).

Blowouts represent a *significant hazard* to human lives in the offshore industry. However, based on the experienced blowouts it *cannot be claimed that blowouts have been a major contributor* to personnel risk in the U.S. GoM OCS and the North Sea.

Nineteen persons died due to H_2S poisoning (Saudi Arabia), and in another blowout, 16 were killed in a fire (Peru). These are the most severe offshore blowouts, worldwide, in which the blowouts were the direct cause of deaths. In addition, two other blowouts caused several fatalities *indirectly*. One of them was the severe pollution incident in Nigeria, where 180 people died (see page 7). Another tragedy occurred when 37 persons were killed during an evacuation because the lifeboat line snapped (Brazil).

EXPOSURE DATA

Exposure data for the North Sea and the U.S. GoM OCS are not easily accessed. The exposure data used here are derived from various sources and coarse estimates.

The sources used for estimating the exposure time comprise:

- WOAD [73]: presenting estimates of the number of personnel-years on mobile units in the U.S. GoM OCS and the North Sea. The source states that the figures are inaccurate and are based on the manning capacities of the mobile fleet.
- Offshore Accident and Incident Statistics Report 1994 [52]: presenting the estimated workforce in the UK offshore sector. No distinction is made between fixed and mobile installations.
- NPD annual report for 1994 [51]: presenting the total number of working hours on fixed installations for each year from 1976, and the total number of working hours on mobile units for each year from 1989.
- The number of wells drilled and wells in production from the SINTEF Offshore Blowout Database [61].
- Telephone conversation with MMS representative [65].

Combined, the above sources establish exposure data for the various areas. The exposure data derived are shown in Table 2.2. How the above sources were combined is explained below Table 2.2. It should be noted that the exposure data are coarse, and care should be taken if these figures are used as input for other analyses.

Table 2.2
Manning Exposure Data for the North Sea and the U.S. GoM OCS, from January 1980 till January 1994

Area	Main installation type	No. of 10^6 working hours in the period 1980-1994	Data source/assumptions
Norway	Fixed	253	NPD Annual report 1994 [51]
	Mobile	64	NPD Annual report 1994 [51], data for 1989-1993. Data for 1980-1988 are extrapolated by using the relative no. of exploration wells drilled in the period.
	Total	317	Sum of above
UK	Fixed	532	*
	Mobile	219	*
	Total	751	Offshore Accident and Incident Statistics Report 1994 [52], assuming 1,800 hours in a working year
US GoM OCS	Fixed	1,350	Total minus mobile below
	Mobile	244	WOAD [73] assuming 1,800 hours in a working year
	Total	1,594	**
Total	**Fixed**	**2,134**	
	Mobile	**528**	
	Total	**2,662**	

* It is assumed that there is a linear relationship between the number of exploration wells drilled and the working hours on mobile units in Norway and the UK. The estimated number of working hours for mobile units in the UK sector will then be 64.1 x (1,859 UK wells/ 543 Norwegian wells) = 219.4 million working hours. When adding the Norwegian number, and also regarding that this figure does not include the relatively limited number of wells drilled on the Dutch, Danish and German sectors, the total calculated North Sea frequency will be somewhat higher than the WOAD estimate for the North Sea. For these purposes it is regarded as sufficiently accurate.

** No exposure statistics exist for the U.S. GoM OCS. The figure is based on the following. The complete U.S. GoM OCS was evacuated during a 1995 Hurricane warning. This involved 26,000 people [65]. When assuming that 50% of the people were working and 50% were resting and that the number of employees has been fairly constant since 1980, the total no. of working hours has been estimated as: 26,000 x 0.5 x 24 x 365 x 14 = 1,594 million working hours.

EXPERIENCED FATAL ACCIDENT RATES

Although the exposure data presented in *Exposure Data* on page 11 are not precise and should not be interpreted as accurate, the data may serve as a basis for coarse FAR estimates. The fatality data presented in *Fatalities in Offshore Activities* on page 9 are considered accurate.

The experienced FAR values for the UK, Norway, and U.S. GoM OCS are presented in Table 2.3

Table 2.3
Experienced Overall FAR Values for the UK, Norway, and U.S. GoM OCS, January 1980 - January 1994 (based on Tables 2.1 and 2.2)

Area	Conditions for the FAR calculations	FAR (fatalities per 10^8 working hours)
UK	Total FAR (incl. the Piper Alpha accident)	36.50
	Total FAR (disregarding the Piper Alpha accident)	14.20
	Contribution to total FAR from blowouts	0.13
Norway	Total FAR (incl. the Alexander Kielland accident)	47.30
	Total FAR (disregarding the Alexander Kielland accident)	8.50
	Contribution to total FAR from blowouts	0.32
US GoM OCS	Total FAR	9.00
	Contribution to total FAR from blowouts	1.13
Total	**Total FAR (incl. the Piper A. and Alexander K. accidents)**	**21.30**
	Total FAR (disregarding the Piper A. and Alexander K. accidents)	**10.40**
	Contribution to total FAR from blowouts	**0.75**

The total experienced FAR values in the North Sea are higher than the FAR values in U.S. GoM OCS. If disregarding the Piper Alpha and the Alexander Kielland accidents, it is seen that the UK North Sea experienced FAR is approximately 50% higher than the U.S. GoM OCS and the Norwegian North Sea experienced FAR values.

Offshore drilling activities are known to involve higher accident frequencies than other offshore activities. Relatively more drilling is carried out in the North Sea than in the U.S. GoM OCS. As seen from Table 4.1 on page 39, the total number of wells drilled in the North Sea from 1980-1994 is approximately 37% of the total number of wells drilled in the U.S. GoM OCS during the same period. In Table 4.4 on page 42, the number of well-years in service for the North Sea is approximately 15% of the number of well-years in service for the U.S. GoM OCS. Further, the average time it takes to drill a well in the North Sea is much longer than the average time it takes to drill a well in the U.S. GoM OCS (Table 4.3). This implies that the total U.S. GoM OCS FAR values should be lower than the North Sea FAR values (disregarding Alexander Kielland and Piper Alpha accidents).

However, the sizes of the installations in the U.S. GoM OCS are on average smaller than the North Sea installations. This means that the work carried out by U.S. GoM OCS non-drilling personnel will, in many ways, differ from the work carried out by non-drilling personnel in the North Sea. Whether this different type of work is more accident prone or not is unknown.

The contribution to the total FAR from blowouts is, however, somewhat higher in the U.S. GoM OCS than in the North Sea. Only two blowouts with one fatality each have occurred in the North Sea during the 1980-1994 period. In the U.S. GoM OCS, the five blowouts resulted in six, five, four, two, and one fatalities.

Since the type of work performed by the majority of the work force is different in the U.S. GoM OCS and the North Sea, a comparison of the North Sea and the U.S. GoM OCS based on these data has therefore limited value.

The SINTEF Offshore Blowout Database shows that 12 of the 20 fatalities occurred on mobile units, and eight occurred on fixed installations. Ten of these 12 fatalities occurred in the U.S. GoM OCS, and two in the North Sea. By using the WOAD exposure data for mobile units [73], the FAR values for experienced blowouts on mobile units in the U.S. GoM OCS and the North Sea are derived and are shown in Table 2.4.

Table 2.4
FAR Values from Blowouts Experienced on Mobile Units in the U.S. GoM OCS and the North Sea

Area	FAR (fatalities per 10^8 working hours)
US GoM OCS	4.1
North Sea	0.8

The reason for the relatively large contribution to the total FAR from mobile units in the U.S. GoM OCS compared with the North Sea is not known. Random variations in blowout consequences may have influenced the difference. It also seems likely that North Sea mobile units are, on the average, of higher quality than U.S. GoM OCS mobile units, partly because of stricter government regulations, partly because of a harsher environment and greater water depths in the North Sea, and partly because U.S. GoM mobile installations, on average, are older than the North Sea installations. Further, North Sea safety standards may be higher than the U.S. GoM OCS safety standards (evacuation possibilities, contingency procedures, contingency training, fire detection, and fire extinguishers, etc.).

COMPARISON WITH OTHER INDUSTRIES

It is of general interest to see what the FAR values are in other industries. The total 1994 FAR value for all industries in the U.S. was 2.6 fatalities per 10^8 working hours [67]. Detailed statistics were unavailable. The total 1992 FAR value for all industries in Norway, including the offshore industry, was also 2.6 fatalities per 10^8 working hours [62]. More detailed FAR values are presented in [43]. These data show that the FAR value for British industry as a whole was four; the chemical industry in Germany, France, and the UK was five. Other FAR values were given, but these data are fairly old. A report concerning fatal accidents in the Nordic countries (Denmark, Finland, Norway, and Sweden) was published in 1993 [20]. Table 2.5 lists the FAR values for the Nordic countries based on that report.

Table 2.5
Experienced FAR Values for the Nordic Countries (1980-1989)

Industry	FAR (fatalities per 10^8 working hours)*
Agriculture, forestry, fishing and hunting	6.1
Raw materials extraction (onshore)	10.5
Industry, manufacturing	2.0
Electric, gas, and water supply	5.0
Building and construction	5.0
Trade, restaurant and hotel business	1.1
Transport, post, and telecommunication	3.5
Banking and insurance	0.7
Private and public services defense, etc.	0.6
Total	**2.0**

*assumed that one employment year includes 1,800 working hours

It is interesting that the FAR values for North Sea activities are far higher than the most risky of the other industries when the Piper Alpha and Alexander Kielland accidents are included. If these two incidents are ignored, the FAR value for the North Sea is approximately at the same level as the most risky land-based industry.

CHAPTER THREE

Blowout Barriers and Analyses

Maintaining well control is important during all well operational phases. Failure to maintain well control will result in a blowout, which may cause severe damage to material assets, the environment, and loss of human lives.

Blowout definition

An uncontrolled flow of fluids from a wellhead or wellbore is classified as a *blowout*. Unless otherwise specified, a flow from a flowline is not considered a blowout as long as the wellhead control valves can be activated. If the wellhead control valves become inoperative, the flow is classified as a blowout [1].

BARRIERS IN WELL OPERATIONS

Barrier definition

A well barrier is an item that, by itself, prevents flow of the well reservoir fluids from the reservoir to the atmosphere [64].

Barrier requirements

The NPD [3] established the following regulations related to well barriers:

- During drilling and well activities at least two independent and tested barriers shall, as a rule, be available in order to prevent an unintentional flow from the well.
- The barriers shall be designed so as to enable rapid re-establishment of a lost barrier.
- In the event of a barrier failure, immediate measures shall be taken in order to maintain an adequate safety level until at least two independent barriers have been restored. No activities for any purposes other than re-establishing two barriers shall be carried out in the well.
- The barriers shall be defined and criteria for failure shall be determined. The position/status of the barriers shall be known at all times.
- The operator shall stipulate requirements to accessibility for the different barriers, and shall be able to provide documentation to show that the requirements have been complied with.
- It shall be possible to test the barriers. Testing methods and intervals shall be determined. To the extent possible, the barriers shall be tested in the direction of flow.

Separate regulations are issued by the NPD for the handling of shallow gas, which include the gas diversion possibility as a second barrier when drilling the tophole section. However, this is not a barrier, according to the barrier definition on page 17. Separate guidelines with supplementary information to the well barrier regulations have been issued by the NPD.

Neither U.S. GoM OCS nor UK regulations include such strict and formal regulations regarding well barriers as the Norwegian regulations do. However, in the UK, a Health and Safety Executive (HSE) guideline previously recommended two barriers during completion and workover activities. UK regulations are, however, now less specific than previous regulations. Otherwise, normal operational practices are followed. In the U.S. GoM OCS, the term barrier is normally not used in association with well operations, and the requirements are not directed towards the number of barriers, but rather are related to the operational aspects [15].

However, in general, the two barrier principle is followed both in the UK and the U.S. GoM OCS even though this is not explicitly stated in the regulations.

The two independent barriers are usually referred to as the *primary* and the *secondary* barriers. For operations in a *killed* well, the hydrostatic pressure is regarded as the primary barrier, and the topside equipment, normally a BOP, is regarded as the secondary barrier. In a *flowing* well, the barriers closest to the reservoir are usually regarded as the primary barrier. This would typically be the packer that seals off the annulus, the tubing below the SCSSV, and the SCSSV. The secondary barriers would then be the tubing above the SCSSV, the Christmas tree main flow side, the casing/wellhead, and the annulus side of the Christmas tree.

Table 3.1 presents the various barrier types. They are grouped according to their functions, how they are operated, and how barrier failures are observed. The barriers listed in Table 3.1 are only examples; several other barriers exist.

Table 3.1
Some Typical Well Barriers

Barrier type	Description	Example
Operational barrier	A barrier that functions while the operation is carried out. A barrier failure will be observed when it occurs.	Drilling mud, stuffing box
Active barrier (Standby barriers)	An external action is required to activate the barrier. Barrier failures are normally observed during regular testing.	BOP, Christmas tree, SCSSV
Passive barrier	A barrier in place that functions continuously without any external action.	Casing, tubing, kill fluid, well packer
Conditional barrier	A barrier that is either not always in place or not always capable of functioning as a barrier.	Stabbing valve (WR-SCSSV)

WELL BARRIER ANALYSIS

Well barrier analyses have become more common in Norway during the last decade. These analyses are performed to:

- compare different well completion alternatives with respect to blowout probabilities
- evaluate the blowout risk for specific well arrangements
- identify potential barrier problems in specific well completions
- assess the effect of various risk reduction measures
- identify potential barrier problems during well interventions

There are typically two main types of barrier:

- Static barriers
- Dynamic barriers

A *static barrier* is a barrier that is available over a "long" period of time. This situation applies during production/injection or when the well is temporary closed in.

A *dynamic barrier* is a barrier that varies over time. This will apply for drilling, workover, and completion operations.

For static barriers, barrier diagrams may be used to illustrate and analyze the relationships between the barriers and the conduits (see Figure 3.2, which illustrates the barriers in Figure 3.1). Barrier diagrams are discussed in more detail in *Barrier Diagrams* on page 22.

Barrier diagrams are normally *not* used for illustrating *dynamic barriers*. For such situations, other methods may be used. Such analyses are typically performed by reviewing *each step* of the *operational procedures*. For each step, the changes in barriers and hazards are identified.

The *Driller's HAZOP* (Hazard and Operability Analysis) is a method for analyzing hazards and operability problems during drilling operations. The method is described by Corner et al. [16]. The method is based on the HAZOP, which was developed for hazard and operability analysis during design of process plants [25].

The Driller's HAZOP follows each step of the operation. Guide words identify hazards and operability problems associated with each step. This method requires an analysis team comprised of personnel with varying fields of competence. The Driller's HAZOP has been further developed and used for well operations other than drilling. It is now frequently referred to as the more general *Procedural HAZOP*. The procedural HAZOP will also include operability problems in addition to the barrier problems.

The author developed and utilized a simpler method for analyzing well barriers in a dynamic barrier situation [34]. This method is also related to the sequence of operations. It is important that the analyst knows the operation well and receives input from the operational personnel as well. A special worksheet with the following entries may be used:

- Procedural step number
- Operational details (briefly describes the operational details included in the procedural step)
- Primary barriers present
- Secondary barriers present
- Evaluation of hazards (typically searching for all incidents that may occur and ruin the barriers)

The main objective of performing the dynamic barrier analyses is to identify steps of the operation where two barriers are not present, or to identify combinations of unreliable barriers. The worksheet may vary from analysis to analysis. This method requires less resources than a procedural HAZOP. The method cannot, however, replace the procedural HAZOP. Typically, the procedural HAZOP is carried out when the operational procedure is nearly completed, while the above method is used during the procedure design. The results and considerations from such an analysis will be important input for the procedural HAZOP.

BARRIER DIAGRAMS

Barrier diagrams are best illustrated through examples. The following example is from a predrilled, gravel-packed production well, which was planned to be abandoned for a period of time. Thereafter, the well was to be re-entered, finally completed, and tied back. Barrier analyses were used to investigate the quality of the barrier situation during the abandoned period. Figure 3.1 shows a simplified sketch of the well barriers during the abandonment period.

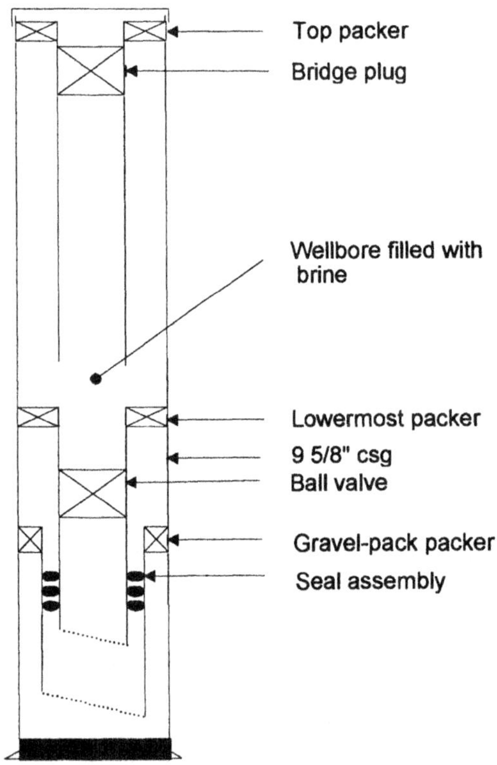

Figure 3.1 Well barriers during the abandonment period for the gravel-packed production well

The main purpose of the barriers is to prevent *reservoir* fluid from entering the *surroundings*. Disregarding the casing, the cement and the formation as barriers, it is seen in Figure 3.1 that the seal assembly, the gravel-pack packer, the ball valve, the lowermost packer, the bridge plug and the top packer are the well barriers. It should be noted that the *brine* is not regarded as an independent barrier, and is therefore excluded from the barrier diagram. This is because if the barriers below the brine start to leak, the brine will leak to the reservoir, and eventually the complete casing volume will be replaced with well fluids.

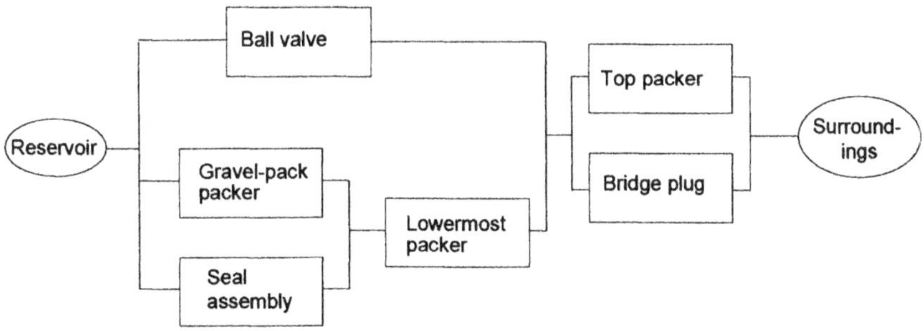

Figure 3.2 Barrier diagram for a temporarily abandoned production well

The possible leakage paths between the reservoir and the environment have to be identified to establish a barrier diagram. The leakage paths in Figure 3.1 will be leakages in the various barriers as shown in Figure 3.2. The relationship between the barriers in Figure 3.1 is illustrated in the barrier diagram shown in Figure 3.2.

It is seen from Figure 3.2 that if the *Ball valve* fails during the abandoned period, the *Top packer* and the *Bridge plug* will be the only remaining barriers between the reservoir and the surroundings. A leakage in any of these two barriers will then result in a leakage/blowout to the surroundings. If, however, the *seal assembly* starts to leak, there will still be a two-barrier situation with the *Lowermost packer*, the *Top packer*, and the *Bridge plug* as barriers. For this branch of the diagram, there is a three-barrier situation.

The complexity of a barrier diagram depends on the complexity of the situation to be analyzed. Barrier diagrams are mostly used when working with different well completion alternatives, but also in connection with subsea Christmas trees and blowout preventer systems.

The probability of a leakage/blowout depends on the leakage probability of each barrier and the structural relationship between the barriers. It is possible to assess the total leakage/blowout probability directly from the structural relationships in the barrier diagram, presupposing that reliability data for the various barriers exist. The complexity of such a calculation will increase with the complexity of the barrier diagram. Therefore, barrier diagrams are frequently transferred to *Fault Trees* for assessing the leakage/blowout probability. This transformation is fairly simple. Figure 3.3 shows the Fault Tree, which represents the barrier diagram in Figure 3.2. Fault Tree Analysis is described in many textbooks [37], and there are several software programs for constructing and analyzing Fault Trees, including the CARA Fault Tree program [14].

The main elements in a Fault Tree are the *TOP event*, the *AND gates*, the *OR gates*, and the *basic events*. The combination of the basic events and the system structure determines whether or not the TOP event will occur. Finding the *minimal cuts* is a crucial element of the Fault Tree analysis. A *minimal cut* represents the combination of barriers that has to fail to experience a leakage/blowout. In Figure 3.3, the combination of the basic event, *leakage in bridge plug,* and the basic event, *leakage in ball valve,* represents a minimal cut, because if these two incidents occur at the same time, the *TOP event*, *leakage/blowout to sea,* will occur.

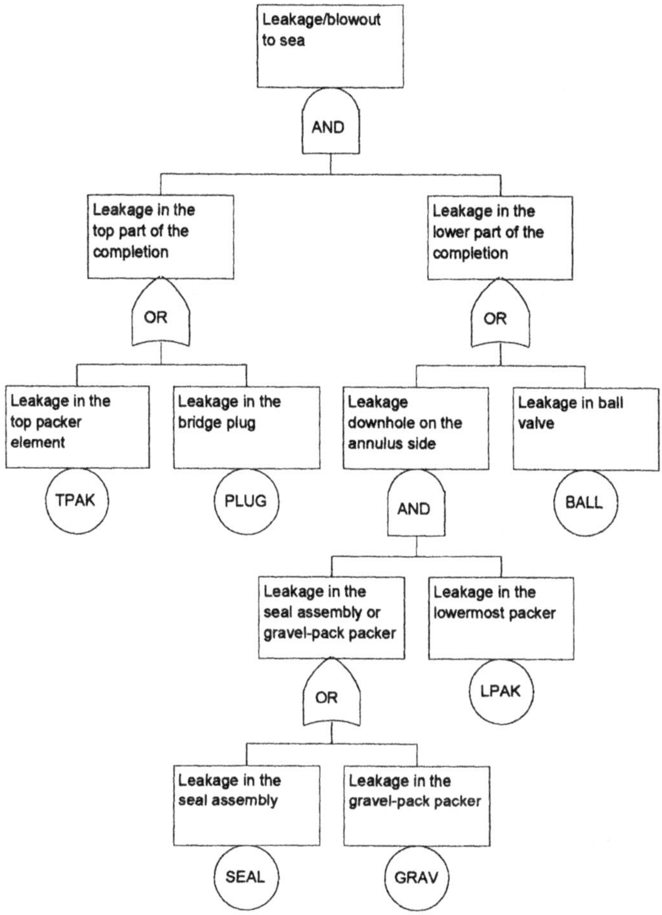

Figure 3.3 Fault Tree representing the barrier diagram shown in Figure 3.2. Source: Cara Fault Tree program [14]

OPERATIONAL PHASES

The taxonomy of the main phases of operation when the blowouts occur is presented in this section. The various phases are selected to avoid comparing blowout causes, frequencies, and consequences in which there are important differences. The distinction between various phases is important when working with risk analyses and/or evaluating risk-reducing measures. One of the main criteria for grouping blowouts according to main operational phases is the blowout barriers present at the various phases. Other criteria are format of exposure data and differences in frequencies experienced in the various phases. The various phases utilized in this book are:

- Exploration drilling blowouts, which are divided into
    - Shallow gas blowouts,
    - "Deep" blowouts
- Development drilling blowouts, which are divided into
    - Shallow gas blowouts,
    - "Deep" blowouts
- Completion blowouts
- Workover blowouts
- Production blowouts
- Wireline blowouts

Exploration drilling is drilling to find hydrocarbons or to determine the extent of a field. When this drilling takes place the *knowledge of the geology and formation is relatively low* compared with development drilling.

Development drilling is drilling of production or injection wells. The *knowledge of the formation is higher* than for exploration drilling.

In principle, drilling a development well is identical to drilling an exploration well. Nevertheless, mainly due to the increased reservoir knowledge, the historical blowout frequency for development drilling is lower than it is for exploration drilling. This is the main reason for making a distinction between development and exploration drilling.

Shallow gas blowouts occur when drilling at shallow depths, and closing in the well with a blowout preventer (BOP) is impossible due to inadequate formation strength, i.e., it is a *single barrier situation.* The single barrier is the hydrostatic pressure from the mud column. "Deep" blowouts obviously occur deeper than shallow gas blowouts. Normally the BOP, the casing, and the formation are the secondary barriers, in addition to the hydrostatic pressure from the mud column, which is the primary barrier.

A *Well completion* includes the steps to prepare a development well for production after drilling is finished. During well completion the barriers will vary depending on the operation. At the outset, drilling barriers typically will be present. During the final stages, the production barriers will be present. In between, various other barriers will be utilized. Plugs may be used in the well and in the wellhead during specific operations. Certain types of equipment will be run into the well, which will disable the barriers for certain periods during the running of various equipment.

A *well workover* is a well maintenance operation. The maintenance operation may be required for technical well problems or productivity problems. A well workover will usually include pulling the production tubing, either partly or fully. During a well workover, the barriers will vary significantly, as in the well completion operation. When starting and ending the operation, typical production barriers will be present. Plugs may be used in the well and in the wellhead during specific operations. Certain types of equipment will be run into the well that will disable the barriers for certain periods.

The *production* phase is the most static barrier situation. Typically, a SCSSV and Christmas tree are the barriers that can be activated to close in the well. In addition, the packer, tubing, casing, and formation are passive barriers.

Usually, a *wireline operation* is well maintenance work. During a wireline operation, a piece of equipment is run on a wireline into the tubing either to make replacements in the well, to install new equipment or to perform well surveys. These operations will disable the SCSSV. A wireline BOP and/or lubricator is then placed on top of the Christmas tree as an extra barrier.

ACCIDENT INVESTIGATION MODELS FOR BLOWOUTS

INTRODUCTION

Accident investigations are carried out to reveal:

- what happened?
- why it happened?
- how similar accidents may be prevented in the future?

From an accident researcher's point of view, the main objective of an accident investigation is to identify measures that reduce the future accident risk. This may be any type of measure regarding technical installations, manning, training of personnel, management, etc. Placing blame for the accident is unimportant.

After an accident, a public inquiry is frequently demanded to find the specific accident's causes. Such investigations are often performed to find someone to blame or prosecute. This approach may motivate persons with knowledge of the accident either to cover up their own mistakes or to cover up a colleague's mistakes. Furthermore, without pinpointing the causes of the accident, protective measures cannot be implemented. Therefore similar incidents will likely occur in the future.

In Norway, accident investigations are usually performed after severe occupational accidents, accidents involving severe material damage or severe pollution incidents. The quality of the accident investigations can, however, in many cases be questioned. The following problems are frequently associated with accident investigations:

- no proper description of the accident exists
- a systematized and clear documentation of findings and conclusions is missing
- the traceability is low

These problems are to a large extent caused by not applying formal accident investigation methods during the investigation. Many accident investigation models or methods have been developed. By applying a model that is suitable for the accident, more thorough accident investigation and reporting may be secured.

In safety research, there has been a focus on developing models of accidents that support inquiries into accidents and design of remedial actions. When establishing/developing accident models, these models are based on accident theory. This theory is an explanation of the causes and effects that lead to the accident. An accident theory may describe causes and relations linked to single incidents, accident patterns, or both.

The various contributing causes to an accident may be related to the following:

- *Physical and technical* aspects such as energy, mass, force, strength, speed, etc.
- *Human* aspects such as awareness, skill, knowledge, motivation, attitudes, etc.
- *Social* aspects such as lifestyle, standard of living, traveling patterns, living patterns, etc.
- *Political and administrative* aspects such as laws and regulations, planning, procedures, operation, maintenance, etc.

Accident causes are often complex because they may be related to technical, human, social, and political levels.

BLOWOUT ACCIDENT INVESTIGATION MODELING

Most blowout accidents have complex causes. The direct cause may often seem simple, but the indirect causes are more complex. For instance, indirect causes may be causes related to inadequate training, inadequate use of personnel, high personnel turnover, low manning, lack of decisions, inadequate preventive maintenance, inadequate procedures, influence from other work, working environment, etc.

Table 3.2 shows a typical blowout sequence and possible contributing factors to the blowout occurrence and blowout consequences. The list is only tentative; important factors may not have been included.

Table 3.2 shows that to investigate a blowout in detail many factors have to be addressed. No accident model will address every aspect. One model may be suitable for modeling the total accident sequence, but for other models it would be better to study specific contributing factors such as management factors, human system interactions, etc.

When using a known accident model for a total blowout investigation, the following is important.

The model should:

- be a sequential type model (Process type)
- include events similar to the events shown in Table 3.2
- address management factors
- address human factors

Table 3.2
Typical Blowout Sequence and Possible Contributing Factors to the Blowout Occurrences and Consequences

Change of state	Accident phase	Factors influencing change of state or consequence within next state

	Normal operational phase (two barrier situation)	• Laws and regulations
		• Platform/rig drilling and safety equipment quality
		• Site specific conditions (geological)
		• Lack of site specific knowledge (geological)
		• Operational procedures (insufficient or faulty)
		• Lack of safety attitudes reflected in
		  • well plan (drilling program and well design) operating procedures (well control and normal operations)
		  • operating personnel attitude and knowledge
		  • administrative personnel attitude
		• Insufficient operator knowledge

Loss of primary barrier		
	Single barrier phase	• Type of primary barrier loss • Secondary barrier quality/reliability • Lack of understanding of situation • Insufficient or faulty well control procedures • Well kick procedures inadequate or not followed
Loss of secondary barrier		
	Initial blowout phase	• Emergency preparedness • Ignition sources • Flow type and volume • Release point • Structure integrity • Environmental aspects • Structure cost • Manning
Structure and personnel exposed to energy		
	Blowout/damage phase	• Vessel available for various assistance • Well integrity • Wellhead integrity • Surface equipment damage • Rig availability for relief well drilling
Re-establish primary and secondary barrier		
	Normalization	

When comparing the events in Table 3.2 with other accident investigation models, it is seen that *MacDonald's* model [26], the *Occupational Accident Research Unit* (OARU) model [41] and the *International Loss Control Institute* (ILCI) model [11] include sequences that are suitable for blowout investigations. A blowout investigation model should be based on one of these three models.

The OARU model is designed primarily for occupational accidents, while the MacDonald and the ILCI models are geared more towards general losses. The ILCI model, a well-known model with much practical experience worldwide, is the basis for the *International Safety Rating System* (ISRS). During the past fifteen years, the OARU model has been used and further developed in occupational accident research in Scandinavia. The MacDonald model is seldom, if ever, used in accident investigations in Scandinavia. The use of the MacDonald model elsewhere in the world is unknown to the author, but is likely to be limited, because it is seldom referred to.

For the ILCI model, there are checklists for system losses, while the OARU model includes checklists that are related to occupational accidents. The ILCI model and the OARU model have many similarities.

It seems that the best suited accident model for analyzing the complete blowout accident sequence is the ILCI accident model.

Detailed investigations of specific parts of a blowout sequence

It could be of interest to analyze specific parts of the accident sequence in more detail.

For specific studies related to management factors, the *Safety Management and Organization Review Technique* (SMORT) [42], or the M-branch of the *Management Oversight and Risk Tree* (MORT) [39] should be used. These methods will ensure a more thorough analysis of the management factors than the ILCI model will provide.

If there is a need for systematizing incidents that have caused a specific accident with respect to time and persons involved, the *Sequentially Timed Events Plotting* (STEP) model [28] may be used.

If the interaction between humans and system is to be focused on, (e.g., the part of the accident related to the operator's ability to observe well kick indications and to react adequately to minimize the influx and the blowout probability), then ergonomic models such as the *Surry Model* [63] may be selected.

CHAPTER FOUR

SINTEF Offshore Blowout Database

INTRODUCTION

The SINTEF Offshore Blowout Database was initiated in 1984. In January 1996 the database included 380 offshore blowouts.

The following companies sponsored the database as of January 1996:

- Statoil
- Saga Petroleum a.s
- BP Norge
- Elf Petroleum Norge A/S
- Norsk Hydro a.s
- Shell International Exploration and Production B.V
- Dovre Safetec AS
- Scandpower

The quality of the database has significantly improved, both with respect to the user interface and the blowout data.

Presently, the database contains 42 different fields describing various parameters related to the blowouts.

Today, the database has a user-friendly interface. Almost any type of search may be performed to select specific blowouts. Search criteria may be established by selecting predefined codes, specific numeric values, specific free text, or any combination of these. The predefined codes are spelled out to ease understanding.

The database program counts and presents the data, satisfying the search criteria. The selected data may be viewed, printed, or copied to separate files for further analysis (e.g. by database or spreadsheet programs).

The structure and the operation of the database are shown in detail in the *Users' manual* for the SINTEF Offshore Blowout Database [61]. The database structure is briefly presented in the next section.

DATABASE STRUCTURE

The SINTEF Offshore Blowout Database program was written in Paradox for Windows, and the raw database file is in the Paradox format.

Worldwide blowout descriptions and overall exposure data from the U.S. GoM OCS and the North Sea (UK and Norway) are included in the database. The exposure data and the blowout description data are not linked. Therefore, blowout frequencies cannot be directly established.

Blowout database taxonomy

The database contains 42 different fields describing each blowout. Nineteen of the 42 fields have pre-defined text codes, some fields are numeric and some are free text. The various fields are grouped in seven groups. The fields included in each group are listed on page 35. More information about the pre-defined text codes of the various fields is presented in the database's *Users' manual* [61].

DATABASE GROUPINGS

Location description

The location description group includes information related to the offshore installation and its location. Seven different fields are used for this purpose. These are *date, country, field, operator, installation name, installation type*, and *water depth.*

Well description

The well description group includes information related to the well conditions when the blowout occurred. Seven different fields are used for this purpose. These are *well status, well depth, mud weight, casing depth* (last casing), *last casing size, bottom hole pressure,* and *shut in pressure.*

Formation

The formation fields include *rock type, formation age,* and the *local name of the formation.* These fields include little information and are mainly intended for future purposes.

Present operation

Included in the present operation group are three fields. They are *phase* (exploration drilling, development drilling, workover, etc.), *operation* presently carried out (e.g., casing running) and present *activity* (e.g., cementing)

Blowout causes

Blowout causes include four fields. They are *external cause* (stating if an external cause contributed to the incident), *loss of the primary barrier, loss of the secondary barrier* (describing how the primary and the secondary barriers were lost), and *human error.* The field regarding human error, in general, holds low quality information. Human errors are frequently masked.

Blowout characteristics

The 12 fields included in the blowout characteristics group are *flow path, release point, flow medium, flowrate* (low quality), *ignition type, time to ignition, lost production* (low quality), *duration, fatalities, consequence class, material loss,* and *pollution.*

Other fields

The *control method* field describes how the well control was regained. The *remarks* field includes a description of the incident. The *data quality* field includes an evaluation of the source data quality.

EXPOSURE DATA

The exposure data in the SINTEF Offshore Blowout Database are presented as text tables. No direct link exists between the event database and the exposure database. The exposure data cover the period from January 1980 till January 1994.

Drilling exposure data for both development and exploration drilling are included for the following countries:

- US GoM OCS
- Norway
- UK
- The Netherlands (not included in this book)

The data format is number of wells drilled each year within the various categories.

Production exposure data are included for:

- US GoM OCS
- Norway
- UK

Exposure data during production is presented as number of well-years in service.

The format of the exposure data varies between the different areas (Norway, UK, The Netherlands, and U.S. GoM OCS). This is because the various sources present the exposure data differently.

For the production data, the source presents the U.S. GoM OCS data as either flowing oil producers, artificially lifted oil producers, or gas/condensate producers. UK wells in operation are presented as producing oil wells, producing gas/condensate wells, gas injection wells, water injection wells, observation wells, or other wells. For Norway the wells are presented as oil producers, condensate producers, gas producers, or suspended/closed in wells.

The number of completions carried out each year is estimated and is based on the total number of development wells drilled in the North Sea. For the U.S. GoM OCS area, the number of completions is listed by the source.

No overall exposure data exist for workovers and wireline operations.

Some sources presented the exposure data so it could be directly transferred into the database; whereas, for other sources some assumptions and approximations were needed to obtain the required data. The following presents the various sources that were used to obtain the drilling and production exposure data.

Drilling exposure data sources

In May 1996, the U.S. Minerals Management Service (MMS) made computerized lists of all wells drilled in the U.S. GoM OCS available through their site on the Internet [71].

The file includes information on borehole activities such as drilling activity, the number of boreholes completed, and number of shut-ins. Additional information includes the lease number, well name, spud date, well class, surface area/block number, and statistics on well status summary. In total, the datafile includes information about more than 30,000 wells.

The drilling exposure data for UK are based on *Development of the Oil and Gas Resources of the United Kingdom*, Department of Trade and Industry, 1994 [68].

The drilling exposure data for Norway are based on the *NPD Annual reports* from 1980-1994 [51] and computerized borehole lists [13].

The drilling exposure data for The Netherlands are based on *Olie en gas in Nederland opsporing en winning, 1994,* published by the State Supervision of Mines [54].

Production exposure data sources

The production exposure data for U.S. GoM OCS are based on *World Oil's* annual Forecasting issues [74].

The production/injection exposure data for the UK before 1991 are based on well data that were systematically collected in the SINTEF study *Reliability of Well Completion Equipment* [48], and *Development of the Oil and Gas resources of the United Kingdom* [17, 18, 68], *North Sea Field Development Guide* [69], and coarse evaluations where well data were missing. These evaluations were typically based on the number of wells in production during previous or later years, the total number of slots on the platform, the platform installation and removal date, the number of wells completed on the platform each year, and description of drilling and production activities. The data are therefore not exactly correct. The data from 1991, 1992, and 1993 are based on statistics from the Department of Trade and Industry [66].

The production/injection exposure data for the Norwegian sector of the North Sea are from the *NPD Annual Reports* [51].

Drilling exposure data

The overall exposure data for drilling activities in the North Sea and the U.S. GoM OCS are shown in Table 4.1.

Table 4.1
Wells Drilled in the U.S. GoM OCS and the North Sea

Year	US GoM OCS drilling		North Sea drilling		Total	
	Expl.	Dev.	Expl.	Dev.	Expl.	Dev.
1980	347	753	90	149	437	902
1981	328	812	113	153	441	965
1982	371	793	160	140	531	933
1983	370	723	168	118	538	841
1984	537	729	229	141	766	870
1985	484	621	207	180	691	801
1986	253	400	149	132	402	532
1987	387	413	168	172	555	585
1988	502	422	189	216	691	638
1989	417	475	195	209	612	684
1990	463	497	242	168	705	665
1991	292	344	213	189	505	533
1992	179	270	156	226	335	496
1993	300	508	123	239	423	747
Total	**5,230**	**7,760**	**2,402**	**2,432**	**7,632**	**10,192**

As seen from Table 4.1, the total number of wells drilled in the North Sea from January 1980 till January 1994 represents approximately 37% of the total number of wells drilled in the U.S. GoM OCS during the same period.

It should, however, be noted that the drilling period for many of the U.S. GoM wells is of very short duration. If looking at all the wells,

- 1,239 of the 7,760 development wells were drilled in less than 10 days, and
- 834 of the 5,230 exploration wells were drilled in less than 10 days.

The time to drill a well in the Norwegian sector of the North Sea is on the average far higher than the time to drill a U.S. GoM OCS well. In Table 4.2, the average drilling times are shown for Norwegian and U.S. GoM OCS exploration and development wells. Table 4.2 presents the average drilling time for all wells and the average drilling time for all wells drilled in less than 200 days.

Table 4.2
Average Drilling Time for Norwegian and U.S. GoM OCS wells *

	US GoM OCS		Norway	
	All wells (days)	Wells with duration < 200 days	All wells (days)	Wells with duration < 200 days
Development drilling	36.6	32.6	102.4	66.9
Exploration drilling	20.1	14.1	84.5	78.7

* From [71] and [13]

As seen from Table 4.2, the time to drill an exploration well in the Norwegian sector of the North Sea is approximately five times higher than the time to drill a U.S. GoM OCS exploration well. For development wells the relationship is approximately two. These figures will never be exact, because some wells have been temporary abandoned for a period. This book focuses on the blowout frequency related to the number of wells drilled. The time it takes to drill a well is not considered.

Thirteen percent of the U.S. GoM wells are sidetracked wells. Relatively few wells in the North Sea are sidetracked. The wells in the U.S. GoM are primarily sidetracked to deflect the direction of the borehole and encounter an alternate target horizon or potential productive interval at a selected aerial location. Deviation of a wellbore to bypass junk in the hole is not classified as a sidetrack.

A number of wells in the U.S. GoM OCS is completed in producing intervals at subsea depths between 305 m and 3,050 m (1,000 ft and 10,000 ft). In areas where the geology and formation pressures have been previously established, development wells are routinely drilled in 1 to 10 days, due to the unconsolidated nature of the formations at depths above 3,050 m (10,000 ft).

Well completion exposure data

Well completion exposure data are presented in Table 4.3. These data are derived from the drilling data.

Table 4.3
Completed Wells in U.S. GoM OCS and the North Sea

Year	US GoM OCS * Completed development wells	US GoM OCS * Completed exploration wells	North Sea **	Total no. of completed wells
1980	411	36	149	596
1981	454	36	153	643
1982	472	44	140	656
1983	448	47	118	613
1984	459	72	141	672
1985	366	70	180	616
1986	236	26	132	394
1987	256	75	172	503
1988	246	110	216	572
1989	323	91	209	623
1990	332	103	168	603
1991	229	70	189	488
1992	211	52	226	489
1993	394	84	239	717
Sum	**4,837**	**916**	**2,432**	**8,185**

* From [71]
** It is assumed that all North Sea development wells are completed and no exploration wells are completed

Production exposure data

The overall exposure data for production activities in the North Sea and the U.S. GoM OCS are shown in Table 4.4.

In Table 4.4, it is seen that the number of well-years in service for the North Sea represents approximately 15% of the number of well-years in service for the U.S. GoM OCS. The number of wells in service in the U.S. GoM OCS has remained approximately constant during the period, while there were approximately 2.5 times more production wells in the North Sea in 1993 than in 1980.

Table 4.4
Overall Production Exposure Data for U.S. GOM OCS and North Sea

Year	No. of production wells in service at the end of each year								
	US GoM OCS			North Sea			Total		
	Oil	Gas/-cond	Total	Oil	Gas/-cond	Total	Oil	Gas/-cond	Total
1980	3568	3370	6938	375	325	700	3943	3695	7638
1981	4156	3887	8043	407	334	741	4563	4221	8784
1982	4569	4179	8748	463	338	801	5032	4517	9549
1983	4177	3856	8033	512	343	855	4689	4199	8888
1984	3786	3534	7320	576	358	934	4362	3892	8254
1985	3981	3462	7443	601	388	989	4582	3850	8432
1986	4077	3386	7463	685	424	1109	4762	3810	8572
1987	4128	3200	7328	734	456	1190	4862	3656	8518
1988	4161	3283	7444	751	498	1249	4912	3781	8693
1989	3497	3109	6606	810	505	1315	4307	3614	7921
1990	4078	3574	7652	857	538	1395	4935	4112	9047
1991	4158	3485	7643	936	583	1519	5094	4068	9162
1992	4243	3400	7643	1013	606	1619	5256	4006	9262
1993	4351	3512	7863	1117	599	1716	5468	4111	9579
Total	**56,930**	**49,237**	**106,167**	**9,837**	**6,295**	**16,132**	**66,767**	**55,532**	**122,299**

QUALITY OF BLOWOUT DATA

The blowout information fed into the database has various origins. The best blowout descriptions are from blowout investigation reports (public, company, or insurance reports), while the blowout descriptions with the lowest quality are from small notices in magazines. Even in the investigation reports, several crucial facts may be missing, like cause of kick, ignition source, and ongoing activity. This means that the information in these data fields is not specifically stated in the sources.

In total, 380 blowouts from 1957 until November 1995 are included, whereof 317 are from 1970 and later. Table 4.5 presents an overview of the source data quality for all blowouts since January 1970. The criteria for the data quality evaluation are listed in the *Users' manual* for the SINTEF Offshore Blowout Database [61].

Table 4.5
Quality of the Source Data in the Blowout Database

Data quality	US GoM OCS Norway and UK 1970-1979	1980 -1995	Rest of the world 1970-1979	1980 -1995	Total 1970 -1995
Very good	8	29	3	4	44
Good	7	22	3	4	36
Fair	17	36	3	14	70
Low	26	31	15	21	93
Very low	4	6	28	34	74
Total	**62**	**126**	**52**	**77**	**317**

In general, the oil business would benefit if companies were more open about why blowouts occurred. Identifying means to reduce the blowout probability would then be easier. However, it is the author's opinion that oil companies and drilling contractors dislike that their blowouts become publicly known, because this leads to a bad reputation that may hurt the business. Further, the people directly involved in the well operations when control was lost frequently mask their own and their colleagues' mistakes for various reasons. They may, for example, be afraid of losing their jobs, reducing their further career prospects, or being prosecuted after the incident. These are well-known phenomena from all types of accidents, and have negative influences on future accident prevention.

In general, identifying blowouts that have occurred in the U.S. GoM OCS is easier than identifying those in the North Sea. This is because in the U.S. GoM OCS all severe offshore incidents have to be reported to the MMS. The MMS stores parts of this information in a database system, from which the public has access. Further, the MMS also releases public investigation reports more often. North Sea blowouts are identified from rumors, newspapers, articles, etc.; the key blowout information cannot be found in one or some few places. Most oil companies worldwide are reluctant to distribute internal documents regarding the various blowouts.

The SINTEF Offshore Blowout Database covers most blowouts in the North Sea and the U.S. GoM OCS, but several blowouts from other parts of the world are believed to be missing. Blowouts not included from the North Sea and the U.S. GoM are typically blowouts that have never been reported other than in internal company files. It is likely that several underground blowouts have never been reported. Further, for the North Sea, it is likely that some shallow gas blowouts and other minor blowouts are not included, because they have not been reported in any public sources.

When using the blowout database it is always important to bear in mind that the quality of blowout data is highly variable.

OTHER BLOWOUT DATABASES

Three other blowout databases are:

- the WOAD (World Offshore Accident Databank) contains approximately 300 offshore blowouts, in addition to several other types of offshore accidents. WOAD is operated by DNV (Det Norske Veritas Industry Norge AS, Oslo, Norway);
- the ERCB (Energy Resources Conservation Board, Alberta, Canada) database includes nearly 400 onshore blowouts; and
- the NAF (Neal Adams Firefighters, Houston, Texas, USA) database contains nearly 1,000 offshore and onshore blowouts.

Otherwise, MMS stores information in a database about accidents associated with oil and gas operations on the outer continental shelf. The accidents are classified as blowouts, explosions and fires, pipeline breaks or leaks, significant pollution incidents, and injuries and fatalities. For each of the incidents some crucial parameters are given in addition to a brief description of the incident. Printout from this database system is public information [1, 2]. This information has been important source material for the SINTEF Offshore Blowout Database.

It is likely that the only comprehensive, updated blowout databases are the ones listed above, and the SINTEF Offshore Blowout Database. Other smaller blowout databases may, however, exist.

On page 45, brief descriptions of the WOAD, ERCB, and NAF databases are given.

WOAD

In the WOAD database, blowouts are only one amongst many types of offshore accidents, and so the various fields in the database are therefore not tailored to the blowout incident. In total, the WOAD database includes information from approximately 3,000 various offshore accidents. This database focuses on the consequences of the accident, not the causes.

ERCB

The ERCB onshore blowout database is a public database. It contains Canadian onshore blowouts only. Fifty-six different fields describe each incident. The database gives a fairly good description of the various events.

As of October 1994, the ERCB database contained 384 onshore blowouts, of which 241 are from January 1980 to January 1994. In addition, the database contains 209 minor releases, named blows.

ERCB also keeps track of the number of wells drilled and number of service operations (workovers) carried out, so general exposure data exists for frequency estimates.

NAF

The NAF blowout database is not a public database. Many of the onshore blowouts in the NAF database come from the ERCB database. NAF also has access to the ERCB blowout files in Alberta. In addition, blowout data from Texas and Louisiana, and some random data from other parts of the world are included.

The NAF database includes approximately 1,000 onshore and offshore blowouts. According to NAF, several of the blowouts included in ERCB's database should not be labeled as blowouts, but rather as leakages (e.g., leakage in valves etc.).

The author does not have access to information about the NAF database structure or the blowout information filled in for each record. However, there is reason to believe that the quality of the blowout information in this database is at least as high as the ERCB database, since NAF primarily works with blowouts.

Exposure data for blowouts do not exist in the NAF database.

CHAPTER FIVE

Overview of Blowout Data

WHEN AND WHERE DO BLOWOUTS OCCUR?

This section presents an overview of the blowouts that are included in the SINTEF Offshore Blowout Database. The SINTEF Offshore Blowout Database includes 380 offshore blowouts worldwide (at the time of printing), and is constantly being updated. Offshore blowouts from as far back as 1957 are included. Of these 380 blowouts, 124 were experienced in the period from January 1980 till January 1994 in the U.S. GoM OCS and the North Sea (Norwegian and UK waters). Unless otherwise stated, all the data presented in the book are from these 124 blowouts. Table 5.1 shows the number of blowouts in the above mentioned areas that occurred during the different *operational phases* (operational phases are explained on page 25). Table 5.2 presents the number of blowouts that occurred during the different *operational phases* and the *years they occurred.*

Table 5.1
Number of Blowouts During Operational Phases and Areas

Phase	Exploration drilling		Development drilling		Com-pletion	Work-over	Prod-uction	Wire-line	Un-known	Total
Area	Shallow	Deep	Shallow	Deep						
Norway	7	5	1	-	-	1	-	-	-	14
UK	2	2	2	1	-	-	2 (1)*	-	-	9
US GoM OCS	20	11	20	11	7	18	10 (5)*	3	1	101
Total	**29**	**18**	**23**	**12**	**7**	**19**	**12 (6)***	**3**	**1**	**124**

*The figures in parentheses do not include blowouts "caused" by external forces.

Table 5.2
Number of Blowouts During Operational Phases and Years

Year	Exploration drilling		Development drilling		Com-pletion	Work-over	Prod-uction	Wire-line	Unkn-own	Total
	Shallow	Deep	Shallow	Deep						
1980	2	2	1	-	1	-	2	1	-	9
1981	2	1	-	1	5	2	-		-	11
1982	1	-	4	2	-	3	-	1	-	11
1983	4	1	5	3	-	1	-	1	-	15
1984	4	2	1	-	-	1	-	-	-	8
1985	4	2	1	2	-	2	-	-	-	11
1986	-	-	1	-	-	1	-	-	-	2
1987	1	1	-	1	1	1	3(2)*	-	-	8(7)*
1988	-	3	-	1	-	1	1(0)*	-	-	6(5)*
1989	4	2	3	1	-	3	3(2)*	-	-	16(15)*
1990	4	-	2	1	-	3	-	-	1	11
1991	1	3	3	-	-	1	-	-	-	8
1992	-	-	1	-	-	-	3(0)*	-	-	4(1)*
1993	2	1	1	-	-	-		-	-	4
Total	**29**	**18**	**23**	**12**	**7**	**19**	**12 (6)***	**3**	**1**	**124 (118)***

*The figures in parentheses do not include blowouts "caused" by external forces.

The unknown blowout in Tables 5.1 and 5.2 stems from a company overview and job summary from the Neal Adams Firefighters, Revision, 22 April 1991. MMS did not report it. The source states that well no. 2 in this block suffered an underground blowout in March 1990 (an exact date was not specified). The well flowied via an unsealed fault plane from a deeper and higher pressured zone. The well was successfully killed and drilling was resumed. A large diameter kill packer was used in conjunction with two large volume barite plugs. This blowout is not included in the following section, because it is unknown during which phase it occurred.

Otherwise, Tables 5.1 and 5.2 show that most blowouts occur during drilling. It is also important to note that shallow gas blowouts are far more frequent than "deep" blowouts. Blowouts not regarded as shallow gas blowouts are regarded as "deep" blowouts. Criteria for categorizing a blowout for a shallow gas blowout are presented in *Shallow Gas Blowouts* on page 59. It should further be noted that workover blowouts have occurred more often than "deep" development drilling blowouts.

TRENDS IN BLOWOUT OCCURRENCES

Figures 5.1 and 5.2 show the *no. of wells to blowouts* (NWTB) for exploration and development drilling.

The trend analyses in this book use the following methods:

- Laplace test [38]
- The Military Handbook test (MIL-HDBK) [38]
- Regression analysis [40]

The SINTEF-developed PC-program, ROCOF [57], was used for trend testing by the Laplace test and The Military Handbook test. Borland's spreadsheet program, Quattro Pro version 6.0, was used for the regression analyses.

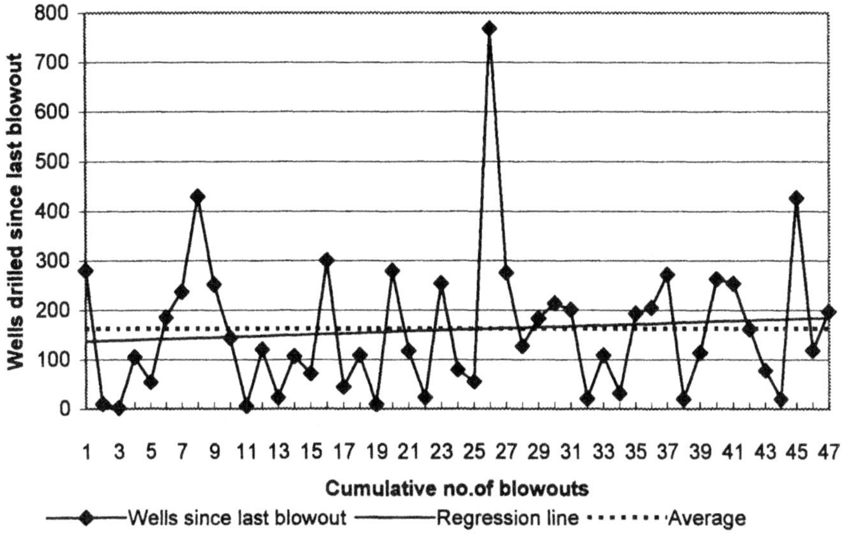

Figure 5.1 No. of wells to blowouts (NWTB) for exploration drilling, shallow gas and "deep" blowouts, the associated regression line, and the average line

There is no significant trend in the number of wells to blowouts (NWTB) for either exploration or development drilling. This was verified through trend tests. On average, a blowout occurred every 162 wells for exploration drilling and every 291 wells for development drilling.

Trends in blowout occurrences related to the main operational phases are investigated in Chapters 6-11, which conclude that only completion operations show a statistically significant trend in blowout frequency. For completion blowouts, a significant frequency reduction was observed. Hughes et al. [35] investigated the blowout trends in U.S. GoM OCS for the period 1960-1984. They concluded that after 1978 the frequency began to decrease.

More specific trends have also been investigated. Chapters 6-11 include the various trend analyses.

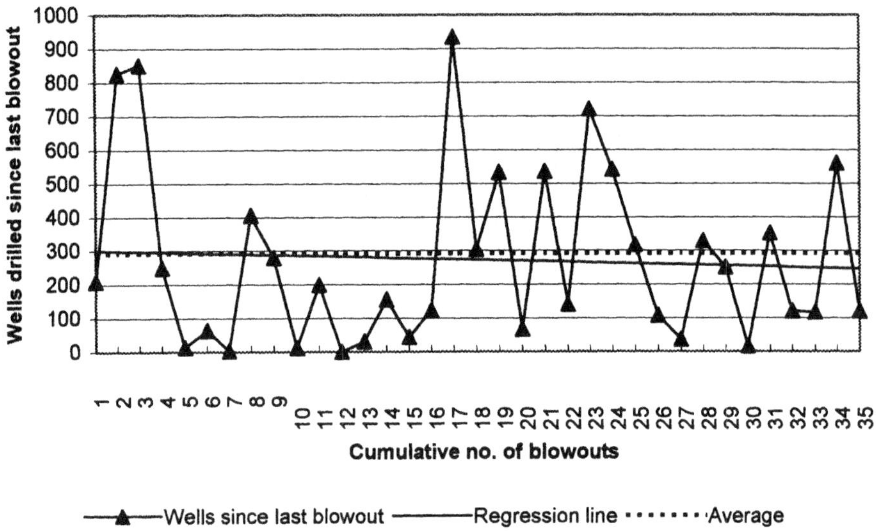

Figure 5.2 No. of wells to blowouts (NWTB) for development drilling, shallow gas and "deep" blowouts, the associated regression line, and the average line

BLOWOUT CAUSES AND CHARACTERISTICS

This section includes an overview of blowout causes and important characteristics. Most of the information is extracted from Chapters 6-10. However, *Ignition Sources and Trends* on page 51 includes an evaluation of ignition causes and trends, which is *not* covered elsewhere in the book.

BLOWOUT CAUSES

Blowout causes are presented in detail in Chapters 6-10. The blowout causes are related to loss of the *primary blowout barrier* and to loss of the *secondary blowout barrier*.

Swabbing and *unexpected high well pressure* are the major contributors to loss of the primary barrier during drilling. However, *loss of hydrostatic pressure while the cement is setting* is also a significant contributor. *Swabbing* is the largest contributor during development drilling, and *unexpected high well pressure* is the biggest contributor to blowouts during exploration drilling.

In the past years, diverter failures have become rare, but prior to that when diverter systems had been used during shallow gas blowout handling, the systems failed more than 50% of the times they were used. A statistical test of the diverter system failures indicated that there is a 85% probability that diverter systems had improved over the period.

The secondary barrier failures during "deep" drilling blowouts are dominated primarily by BOP failures, kelly valve/string safety valve failures, and casing leakages.

For workover blowouts, *swabbing, too low mud weight*, and *trapped gas* were the biggest contributors to loss of the primary barrier. Regarding loss of the secondary barrier during workover blowouts, BOP failures and kelly/string safety valve failures were the dominant contributors.

IGNITION SOURCES AND TRENDS

The loss of material assets and lives is very dependent on whether the blowout ignites or not. Since relatively few blowouts ignite, all blowouts will be in one group in this section. Table 5.3 shows the blowouts which ignited during the various main operational phases.

Table 5.3
Ignition Experience of the Various Blowouts (from Chapters 6-10)

PHASE		No ignition	Immediate ignition < 5 min	Delayed ignition 5 min - 1 hr	1 hr - 6 hr	6 hr - 24 hr	>24 hr	Total
Exploration drilling	Shallow gas	26	-	2	-	1	-	29
	"Deep"	14	1	-	-	3	-	18
Development drilling	Shallow gas	18	3	1	-	1	-	23
	"Deep"	11	1	-	-	-	-	12
Completion		6	-	-	-	1	-	7
Workover		14	2	-	-	-	3	19
Production		6	-	-	-	-	-	6
Wireline		3	-	-	-	-	-	3
Total		**98**	**7**	**3**	-	**6**	**3**	**117**
		83.8%	*6.0%*	*2.6%*		*5.1%*	*2.6%*	

The experience shows that 16% of the blowouts ignited, 6% ignited immediately, 2.5% ignited between five minutes and one hour after the blowout occurred. The remaining 7.5% ignited more than one hour after the blowout occurred.

By reviewing all the blowouts in the database after January 1980, and *excluding the U.S. GoM OCS and the North Sea blowouts,* it is seen that 31 out of these 77 blowouts ignited (i.e., 40% ignited). This high ignition frequency is mainly explained by the fact that blowouts in these areas are identified through public sources alone. These sources do not frequently describe blowouts with small consequences. Therefore, several low consequence blowouts are believed to be missing from the database. Further, the technical standards and procedures may, on average, be of lower quality than the technical standards and procedures for the U.S. GoM OCS and the North Sea areas.

Ignition sources

Reducing the possible ignition sources is an important issue in offshore design and operation. The source material for the 19 ignited blowouts was reviewed. Table 5.4 lists the various proposed ignition sources.

Table 5.4
Overview of Various Proposed Ignition Sources

Degree of certainty	Proposed ignition sources
Certain	Spark from abrasive formation rock that hit derrick metal.
Certain	Traveling block that fell on the drill floor.
Certain	A torch used to cut a hole in the 16-in. casing.
Assumed	The abrasive action at surface.
Assumed	Sparks created by the 1-in. pipe exiting from the well.
Two alternatives	1. Abrasive action in drilling area. 2. Generator no. 3 in engine room. No. 1 was assumed most likely
Two alternatives	1. Static electricity. 2. Sparking due to sand flowing out of the well and cutting metal
Six alternatives	1. Metal grating from gumbo box striking against metal, 2. Mud and debris striking against fluorescent lights in area of gumbo box, 3. Drill cuttings striking against metal, 4. Coffee pot in drillers' office, 5. Static electricity, 6. Electric junction box located over gumbo box
Unknown	No proposed ignition source for 11 blowouts.

In general, ignition sources remain unknown after a blowout. Of the 19 ignited blowouts, only three ignition sources were certain, two were likely or assumed, three included several possible ignition sources, and 11 were unknown.

Ignition trends

Reducing possible ignition sources, both during design and operation, has been focused on in the 1980s and 1990s. Further, during the past few years, shallow gas blowouts have been successfully diverted, compared to earlier years when diverter systems frequently failed. When the well is diverted successfully, the ignition probability is low. Other causes for reduced ignition probability in 1980-1994 may also exist. Based on the above, there is reason to assume that blowouts are not as likely to ignite today as they were in the early 1980s. Various analyses regarding blowout ignition trends were performed to verify if this focus has caused a significantly decreasing ignition trend.

Figure 5.3 shows *the number of blowouts* to ignition for all blowouts, not including blowouts with external causes.

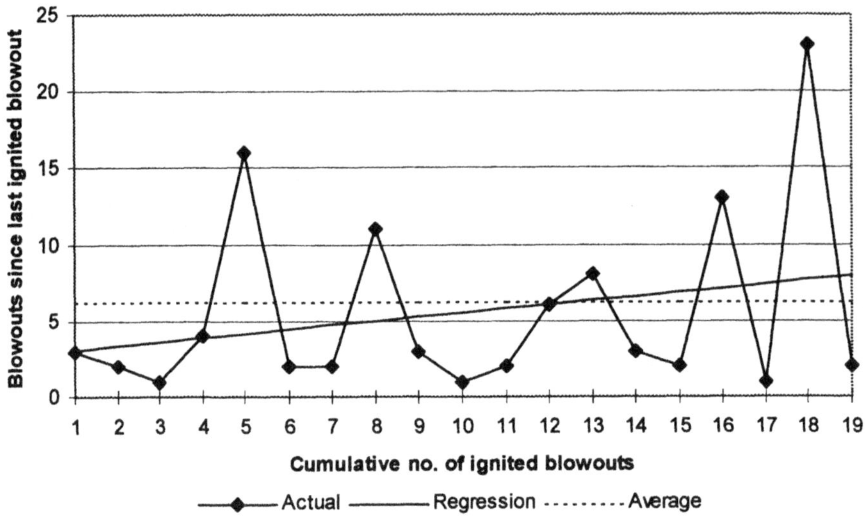

Figure 5.3 No. of blowouts to ignition is shown for all blowouts, not including blowouts with external causes, with the associated regression line and average line

The regression analysis shows that the probability of a decreasing trend is 70% (i.e., it cannot be regarded as a significantly decreasing trend). It is here, however, important to note that the regression analysis does not include censored data. The last ignited blowout was blowout number 105 of the total 118 blowouts. This means that the last 13 unignited blowouts are not considered in the regression trend analysis.

The ignited blowouts and the cumulative number of blowouts also have been analyzed with respect to trends by using the Laplace test [38] and the MIL-HDBK test [38] for cumulative data sets. These two tests confirmed that there was a significantly decreasing trend. The significance probabilities for a decreasing ignition trend for the Laplace test and the MIL-HDBK test were 94% and 89%, respectively.

The ignition frequency related to the cumulative number of blowouts may be calculated by the following expressions:

Power law model: $\alpha\beta t^{(\beta-1)}$
where $\alpha = 0.459$ and $\beta = 0.779$

Log linear model: $EXP(\alpha+\beta t)$
where $\alpha = -1.263$ and $\beta = -0.011$

t denotes the cumulative number of blowouts. The results from the blowout frequency estimation based on the two above expressions for the interval 1-118 blowouts are shown in Figure 5.4.

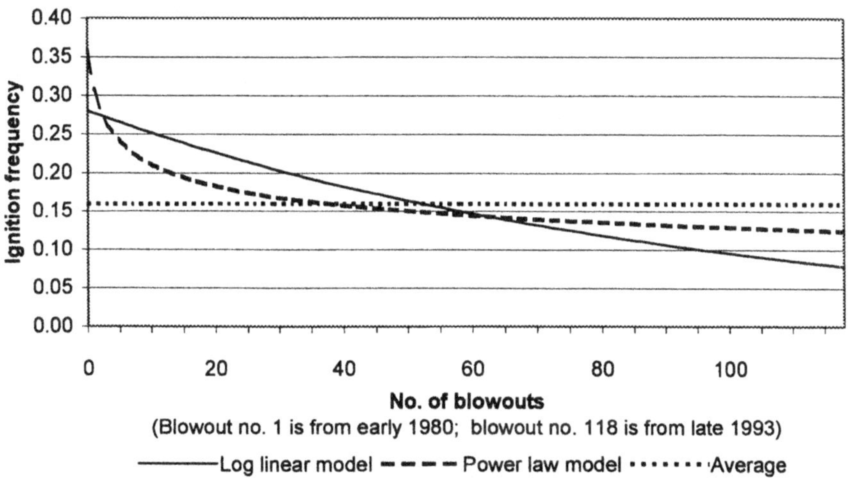

Figure 5.4 Estimated ignition frequency vs. cumulative no. of blowouts

Blowout 118 represents the 1993/1994 ignition frequency level. The estimated frequency for 1993/1994, based on the two different estimation methods, will then be:

Ignition probability = 0.079 per blowout (Log linear model)
Ignition probability = 0.124 per blowout (Power law model)

The identified significant trend in blowout ignition frequency should be credited in risk analyses. Both the Log linear and the Power law models fit fairly well, but not perfectly, to the ignition data. *It is recommended that an average ignition probability of 0.10 be used as an overall input figure for risk analyses.*

BLOWOUT POLLUTION

None of the blowouts caused severe pollution. Experience from other periods and areas do, however, show that blowouts may cause severe pollution. Nevertheless, the probability of severe pollution is low. More specific information related to pollution from blowout incidents is presented for each of the main operational phases in Chapters 6-10 and in Chapter 2.

BLOWOUT DURATION

Table 5.5 presents the duration of the various blowouts.

Table 5.5
Duration of the Various Blowouts

PHASE		< 10 min	10 min - 40 min	40 min - 2 hr	2 hr - 12 hr	12 hr - 5 days	> 5 days	Un-known	Total
Exploration drilling	Shallow	-	2	2	5	9	6	5	29
	"Deep"	1	1	1	2	7	2	4	18
Development drilling	Shallow	3	1	4	2	6	2	5	23
	"Deep"	1	-	-	3	4	3	1	12
Completion		1	-	-	-	4	2	-	7
Workover		3	2	-	1	7	3	3	19
Production		-	-	-	-	5	-	1	6
Wireline		-	1	-	1	1	-	-	3
Total		**9**	**7**	**7**	**14**	**43**	**18**	**19**	**117**
		7.7%	**6.0%**	**6.0%**	**12.0%**	**36.8%**	**15.4%**	**16.2%**	

If assuming that the blowouts with *unknown duration* have the same duration distribution as the blowouts with *known duration,* the following may be claimed: 16% of all the blowouts lasted less than 40 minutes, 21% lasted between 40 minutes and 12 hours, 44% lasted between 12 hours and five days, and 18% lasted more than five days. This distribution of durations is nearly identical to the duration distribution for U.S. GoM OCS blowouts in the period 1960-1984 [35].

BLOWOUT FLOW MEDIUMS

Blowout flow mediums are listed in Table 5.6.

Table 5.6
Blowout Flow Mediums for U.S. GoM OCS and North Sea Blowouts

Flow medium	Expl. drilling	Dev. drilling	Comp-letion	Work-over	Prod-uction	Wire-line	Total	Total grouped
Shallow gas	26	20	-	-	-	-	46	53*
Shallow gas, oil	-	1	-	-	-	-	1	
Shallow gas, water	3	1	-	-	1**	-	5	
Shallow water	-	1	-	-	-	-	1	
Gas (deep)	15	8	6	13	4	2	48	51
Gas (deep) (gas-lift gas)	-	1	-	-	-	-	1	
Gas (deep) (trapped gas)	1	-	-	-	-	-	1	
Gas (deep), water	-	1	-	-	-	-	1	
Condensate, gas (deep)	-	1	-	1	-	-	2	2
Oil	-	-	-	2	-	1	3	12
Oil, gas (deep)	2	1	1	2	1	-	7	
Oil, gas (deep), water	-	-	-	1	-	-	1	
Total	**47**	**35**	**7**	**19**	**6**	**3**	**117**	

* Six of these blowouts were also reported with H_2S

** Stems from a blowout outside the casing from a shallow zone during production

As seen from Table 5.6, gas is the most common blowout flow medium. Oil blowouts are rare. The investigation of U.S. GoM OCS blowouts for the period 1960 until 1984 showed approximately the same blowout flow medium distribution [35]. Blowout flow mediums during drilling and workover are discussed more thoroughly in *"Deep" Drilling Blowout Characteristics* on page 94, and in *Workover Blowout Characteristics* on page 116.

CHAPTER SIX

Drilling Blowouts

Most offshore blowouts occur during well drilling (see page 46, *When and Where do Blowouts Occur?*). Wells are generally classified as:

- *Exploration*
- *Development*

In Norway, *exploration wells* are classified into two groups:

- *Wildcats* (wells drilled in an unproved area)
- *Appraisal wells* (wells drilled following a discovery to determine the extent of an oil field or gas field)

In the UK, only wildcats are regarded as exploration wells, while appraisal wells form a separate category [68]. In the US, an exploratory well may be included in one of five different categories [8]. However, these five categories are included in the wildcats and appraisal wells described above.

This book classifies appraisal wells and wildcats as exploration wells. This is because blowout descriptions commonly do not specify whether the well was a wildcat or an appraisal well.

A *development well* is a well drilled within a proven area of an oil and gas reservoir to a depth of a stratigraphic horizon known to be productive.

Drilling blowouts may occur at nearly all well depths. In some wells, very shallow gas pockets have been observed. In terms of well control, shallow gas blowouts are different from blowouts stemming from deeper zones of the well. The drilling blowouts described in this book are divided in two main types:

- *Shallow gas blowouts*
- *"Deep" blowouts*

All drilling blowouts that are not regarded as shallow gas blowouts are classified as "deep" blowouts.

SHALLOW GAS BLOWOUTS

The general criterion for a shallow gas blowout is a well depth less than 1,200 m (3,900 ft). At such depths the formation fracture gradient generally is low and the well usually is not closed in by a blowout preventer (BOP). Closing in the well might lead to severe cratering. Due to the high deliverability of shallow sands, the large hole diameter, and the short distance from the gas zone to the surface, shallow gas blowouts rapidly progress from influx up to a full blowout.

When drilling at shallow depths, there is normally only one blowout barrier, the drilling fluid. Diverter systems, which lead the gas away from the installation, are installed in most cases. Some bottom-supported drilling platforms are not equipped with diverter systems, when shallow gas is not expected. Floating installations may also be moved away from the well in case of a blowout. In recent years, it has become common for floating installations to drill the shallow sections without a riser, which prevents the gas from being brought directly back to the installation.

When drilling without a riser, the drilling fluid is usually seawater with a density of 1,030 kg/m^3 (8.6 lb./gallon). Mud, which has to be disposed on the seafloor, is usually not used. This limits the hydrostatic pressure in the well, and thereby increases the blowout probability. According to Hellstrand, Statoil's main goal is to avoid having the gas brought to the rig [27]. It is questionable whether this is the best way to drill shallow well sections. According to Grepinet, Total's policy is not to allow a kick to occur (i.e., to keep hydrostatic well control) [23].

Therefore the company recommends the riser option in operations with floating units.

Database criteria for shallow gas blowouts

If *one or more* of the following items of information are clearly indicated in the database *source* file, the blowout is categorized as a shallow gas blowout.

- The well depth is less than 1,200 m (3,900 ft).
- The source states shallow gas as the flow medium.
- Only the conductor casing is run.
- The BOP is not installed on the wellhead.
- The gas flow is diverted, and no attempts are made to close in the well.
- The actual blowout source reservoir is far from the target reservoir.

Therefore, if a blowout included one or two of the above characteristics, but in fact had a deeper depth than 1,200 m (3,900 ft), it could be classified as a shallow gas blowout.

It is not suitable to apply a strict definition of what is to be considered a shallow gas blowout and what is not, in terms of well flow. Incidents involving complete loss of well control are always regarded as shallow gas blowouts. Some shallow gas incidents do, however, include minor releases of gas for a short period when drilling without a riser. These are frequently referred to as shallow gas incidents, not blowouts.

Most shallow gas blowouts described in this book are incidents where well control was completely lost. Some incidents with a limited flow, but of long duration, are also included.

Shallow gas prediction

The exact spud location is frequently based on experience from previous wells drilled nearby and seismic surveys, which helps to avoid drilling into shallow gas pockets. Experience, however, proves that the probability of failing to predict shallow gas pockets has been very high. Østebø et al. investigated 60 exploration wells drilled in four different areas on the Norwegian Continental Shelf from 1978-1986 [75]. They predicted no shallow gas for 31% of these wells, and shallow gas for

69%. For 47% of the wells where shallow gas *was not predicted*, shallow gas was experienced. Further, for 45% of the wells where shallow gas *was predicted*, shallow gas was not experienced. Predicting shallow gas during those years had little or no value.

According to Hellstrand, shallow gas prediction improved significantly after the 1985 West Vanguard accident [27, 72]. All of the shallow gas zones drilled from 1986 until 1990 in five Haltenbanken wells (west of mid Norway) were predicted. Shallow gas was, however, also predicted in some other wells, but not encountered. This improvement in shallow gas prediction is claimed to be the result of using 2D, and in particular, 3D seismic.

West Vanguard blowout, Norwegian sector of the North Sea, 1985
When drilling the shallow section of a well, a shallow gas flow occurred. The gas diverting system failed. The released gas ignited. One person was killed, and the semisubmersible drilling rig experienced severe damages.

According to Moore, the use of 2D and 3D seismic does not guarantee that all shallow gas accumulations will be predicted [50]. The risk of drilling unexpectedly into shallow gas pockets will, however, be reduced. One of the UK Offshore Safety Division's (OSD) aims is to encourage the industry to use 2D and 3D seismic [50].

SHALLOW GAS BLOWOUT EXPERIENCE

The experience presented in this section is based on the SINTEF Offshore Blowout Database (See section *SINTEF Offshore Blowout Database* on page 33) for January 1980 till January 1994 in the U.S. GoM OCS and the North Sea (UK and Norway). A total of 52 shallow gas drilling blowouts has been recorded. Table 6.1 lists the various installation and well types.

Table 6.1
Shallow Gas Blowouts Experienced for Various Installation vs. Main Well Type

Installation type	Development drilling	Exploration drilling	Total
Jacket	19	-	19
Jack-up	2	12	14
Semisubmersible	2	16	18
Unknown	-	1	1
Total	**23**	**29**	**52**

As seen from Table 6.1, approximately the same number of shallow gas blowouts has occurred with jack-ups and semisubmersibles during exploration drilling. Blowout frequencies, according to type of installation, cannot be compared, because the total number of wells drilled with the different types of drilling rigs is unknown. Most of the shallow gas blowouts during development drilling have occurred during drilling from a fixed installation.

Figures 6.1 and 6.2 show the *number of wells to blowout* (NWTB) for shallow gas blowouts, the associated regression lines, and the average lines.

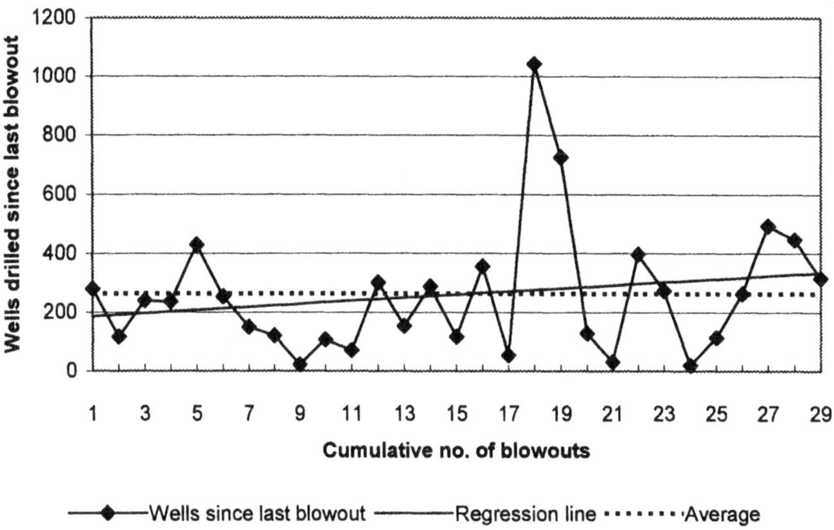

Figure 6.1 Number of wells to blowout for shallow gas blowouts during exploration drilling, the associated regression, and average lines

Figure 6.1 indicates a slight increase in NWTB from 1980 until 1994 for exploration drilling shallow gas blowout occurrences. An overall significant statistical trend in the exploration drilling shallow gas blowouts NWTB could not be concluded by any of the statistical methods used.

Figure 6.2 indicates a slight decrease in NWTB from 1980 until 1994 for development drilling shallow gas blowout occurrences. An overall significant statistical trend in the development drilling shallow gas blowouts NWTB could not be concluded by any of the statistical methods used.

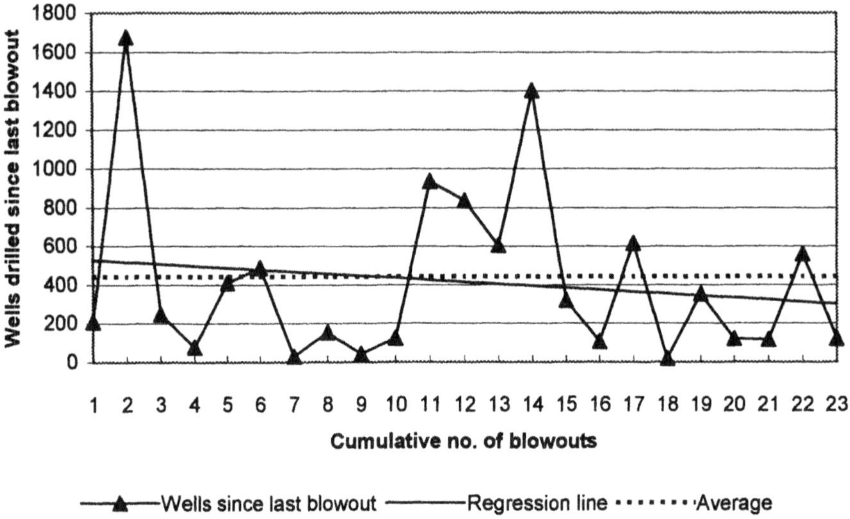

Figure 6.2 No. of wells to blowout for shallow gas blowouts during development drilling, the associated regression, and average lines

During the past decade industry put major emphasis into shallow gas blowout risk reduction means. These means mainly focused on diverter systems, riserless drilling, and handling procedures. Effects of these efforts will be discussed in association with the more detailed analysis to follow.

Table 6.2 shows the operations and activities in progress when the shallow gas drilling blowouts occurred.

Table 6.2
Operations and Activities in Progress when the Shallow Gas Blowouts Occurred

Operation ⇒	Drilling activity		Casing running		Other operations		Total		
Activity ⇓	Dev.	Expl.	Dev.	Expl.	Dev.	Expl.	Dev.	Expl.	Total
Actual drilling	3	12	-	-	-	-	3	12	15
Tripping out	10	5	-	-	-	-	10	5	15
Out of hole	1	1	-	-	-	-	1	1	2
Nipple down BOP	-	-	-	1	-	-	-	1	1
Changing equipment	-	1	-	-	-	-	-	1	1
Waiting on cement	-	-	5	3	-	-	5	3	8
Actual casing running	-	-	1	-	-	-	1	-	1
Install BOP	-	-	-	-	-	1	-	1	1
Milling and reaming	-	-	-	-	-	1	-	1	1
Leak off test	-	-	-	-	-	1	-	1	1
Temporarily plugged	-	-	-	-	1	-	1	-	1
Unknown	1	1	1	-	-	3	2	4	6
Total	**15**	**20**	**7**	**4 (3)***	**1**	**3**	**23**	**30(29)***	**53(52)***

* Figures in parentheses denote number of blowouts. For some blowouts two activities were listed

Seventy percent of all shallow gas blowouts occurred during the following activities: *actual drilling, tripping out*, and *waiting on cement to harden*. Disregarding the blowouts with unknown activities, 80% of all shallow gas blowouts occurred during the following activities: *actual drilling, tripping out*, and *waiting on cement to harden*.

SHALLOW GAS BLOWOUT CAUSES

The likely causes for losing the primary barriers during these operations are discussed in association with Table 6.3. Table 6.3 lists the experienced causes for losing well control.

As seen from Table 6.3, experience shows that the shallow gas blowout frequency is approximately 2.3 times higher during exploration drilling than during development drilling. The listed frequencies are based on all the drilled wells, not only for wells drilled in areas with shallow gas.

Table 6.3
Primary Barrier Failure Causes for Shallow Gas Drilling Blowouts

Primary barrier failure		Development drilling blowouts		Exploration drilling blowouts		Total
		No. of blowouts	Frequency (blowouts/well drilled)	No. of blowouts	Frequency (blowouts/well drilled)	
Too low hydro-static head	unexpected high well pressure	2	0.00020	8	0.00105	10
	too low mud weight	2	0.00020	4	0.00052	6
	swabbing	10	0.00098	5	0.00066	15
	gas cut mud	-	-	2	0.00026	2
	disconnected riser	-	-	1	0.00013	1
	annular losses	2	0.00020	2	0.00026	4
	while cement setting	5	0.00049	2	0.00026	7
	unknown	3	0.00029	3	0.00039	6
Poor cement		1	0.00010	1	0.00013	2
Formation breakdown		-	-	1	0.00013	1
Total		**25 (23)***	**0.00226****	**29**	**0.00380**	**54 (52)***

* Figures in parentheses denote number of blowouts. For some blowouts two primary barrier failures are reported.

* Based on number of blowouts.

As Table 6.3 shows, most primary barrier failures are related to *too low hydrostatic head.* Poor cement and formation breakdown were the causes of the three other primary barrier failures. The experienced reasons for losing the primary barrier will be discussed in the following.

The two blowouts caused by poor cement occurred after the casing operation and during drilling. Only a limited amount of gas was flowing. For the formation breakdown incident, gas was observed at the seafloor after the BOP had been run.

Swabbing

Swabbing is the dominant cause of losing the hydrostatic barrier and hence leading to shallow gas blowouts. Swabbing creates a suction in the wellbore which may induce well fluids out of the formation, creating a kick. Swabbing is usually caused by pulling the drillstring too quickly out of the well.

Approximately 40% of the shallow gas blowouts in development wells and 20% of shallow gas blowouts in exploration wells were caused by swabbing.

In many cases, other factors help create the swabbing effect, or the well has little tolerance with respect to swabbing. For three of these blowouts (in Table 6.3), sticky clay (gumbo) balled up the drill bit and/or the stabilizer, thereby reducing the annular space for mud passage. For one of these incidents, a very narrow hole was reported. Reportedly, one of the blowouts was caused by swabbing when annular losses were experienced after pulling out of the well, which probably reduced the swabbing tolerances. Another blowout was reportedly caused by too low mud weight and swabbing, which also probably reduced the swabbing tolerances. It is important to keep in mind the quality of the blowout data (See *Quality of Blowout Data* on page 42).

Some of these blowouts, reported to be caused by swabbing, may have been caused by improper hole fill-up during tripping to replace the volume of the pipe pulled out.

Swabbing can never be eliminated, but the probability of a blowout may be reduced by taking precautions. The most obvious precautions are:

- Circulate mud while pulling out of the hole (possible when using topdrives).
- Pull out carefully and pay special attention to narrow holes and sticky clay problems.

Generally, it is more likely to experience the swabbing effect in narrow holes. The use of small diameter pilot holes has become normal procedure in the North Sea when drilling the shallow section of a well. NPD also recommends this practice when drilling in areas where shallow gas might be expected [3]. This will thereby increase the swab probability. The pilot holes are used partly to reduce the shallow gas flow, and partly to be able to dynamically kill a well. The technique is based on increasing the annular friction by high rate mud pumping. According to Adams, dynamic killing is not possible due to inherent hole washouts in soft, shallow sediments [5]. Hellstrand also realizes this problem [27], although Statoil has controlled a few shallow gas incidents by using this method, but the company is not sure whether the well was killed by the annulus friction or by the weight of the kill mud.

Unexpected high well pressure/too low mud weight

Unexpected high well pressure and *too low mud weight* are described together because it is likely that an unexpected high well pressure was the reason for reporting too low mud weight, not because the mud weight was mixed to a lower density than specified.

As expected, this blowout cause is far more frequent during exploration drilling than during development drilling, primarily because knowledge about the shallow formation is far better during development drilling.

It is assumed that seismic improvements and increased knowledge of the various areas described in this book have reduced the frequency of such incidents. This assumption was tested by applying the three different statistical trend analyses. These tests, with a confidence limit of approximately 80%, indicated that the blowout frequency decreased for blowouts caused by unexpected high well pressure.

While cement setting

Eight of the blowouts reported in Table 6.3 occurred *while cement was setting.* These blowouts occurred after the casing had been run and cemented. When the cement is in the transition state, i.e., changing from fluid to solid, it will start to stick to the surface of the hole and to the casing. This will reduce the hydrostatic pressure on the formation, and gas may start to flow through or along the cement sheet.

Most such blowouts will be outside the casing, but some occur in the drilling annulus.

Table 6.3 shows that the experienced frequencies for *while cement was setting* before gas blowouts are similar for exploration and development drilling.

A so-called "gas-tight" cement was developed to reduce this problem, but such blowouts still occur. For one of the blowouts, so-called "gas-tight" cement was used. However, this "gas-tight" cement was still in its development stages. Today other additives are used. A statistical test on this type of blowout was carried out for all drilling blowouts [30]. This test indicated that the frequency of this type of blowout actually had increased from 1980-1992. However, recent use of "gas-tight" cement is unknown by the author.

Table 6.3 also includes *poor cement* as a blowout cause. Two blowouts were observed after cementing and during operations. Both of these blowouts included small gas releases.

Annular losses

Four of the shallow gas drilling blowouts were caused by *annular losses*. Annular losses occur when the hydrostatic pressure from the mud column exceeds the formation fracture gradient. This causes fluid to enter the formation, and possibly, the well to kick.

Three of these blowouts occurred during normal drilling operations. One source also listed swabbing as a blowout cause. According to the source, initially annular losses occurred, and thereafter the well was swabbed in when pulling out of the hole. The last of the blowouts occurred during cementing because the cement density was too high.

Gas cut mud

Gas cut mud is a reported cause for two of the exploration well blowouts. Gas cut mud is caused by formation gas, which is mixed with the return mud when drilling through a gas bearing layer. This mixture of gas and mud will reduce the density of the mud, and thereby the hydrostatic pressure. Gas cut mud may also have been a contributing factor for some of the other blowouts. If the mud is gas cut, it will, for instance, be much easier to swab in the well, due to the reduced hydrostatic pressure on the formation.

Other

One of the shallow gas blowouts occurred when *disconnecting the riser* to run a large diameter hole opener. Disconnecting the riser will cause approximately 25 m (80 ft) of the hydrostatic column to be lost (from sea level to the rotary kelly bushing), and decreasing the hydrostatic pressure.

Five shallow gas blowouts, with unknown causes for losing the hydrostatic head, are reported.

The last shallow gas blowout listed in Table 6.3 occurred due to a formation breakdown. After running the BOP, gas bubbles were observed near the wellhead.

SHALLOW GAS BLOWOUT HANDLING

Actions taken to reduce the consequences of the blowout include two main objectives: Avoid exposing the installation to the gas and avoid seafloor cratering.

For non-floating installations, diverter systems are used to lead the gas overboard to avoid damage and danger of an explosion or fire. On some occasions, shallow gas blowouts have been closed in. Closing in a shallow gas blowout greatly increases the chances for a blowout to occur outside the casing, and cratering, which again, in a worst case, may cause a bottom-supported platform to tilt and capsize.

More alternatives are available for floating drilling structures than for bottom-supported drilling installations. These are:

- Diverter systems (topside or subsea)
- Disconnecting riser and pull off
- Drilling without riser

When drilling without a riser, or disconnecting the riser, the hydrostatic pressure from the seawater will reduce the flow. By using subsea diverters the gas may be diverted subsea, and thereby not be brought back to the rig. Subsea diverters have also been used to some extent to close in the wells for short periods (until gas has been observed at the seafloor).

Table 6.4 lists the experience related to shallow gas blowout handling.

Table 6.4
Shallow Gas Drilling Blowout Handling

Shallow gas handling	Development drilling		Exploration drilling		Total
	No. of blowouts	Distribution (%)	No. of blowouts	Distribution (%)	
Failed to stab kelly valve	1	4.0	-	-	1
Failed to close BOP	-	-	1	3.2	1
BOP failed after closure	1	4.0	-	-	1
BOP not in place	1	4.0	1	3.2	2
Diverted, no problem	11	44.0	5	16.1	16
Failed to operate diverter	2	8.0	2	6.5	4
Diverter failed after closure	4	16.0	7	22.6	11
Drilling without riser	-	-	5	16.1	5
Disconnected riser	-	-	1	3.2	1
Fracture at casing shoe	2	8.0	5	16.1	7
Poor cement	1	4.0	-	-	1
Wellhead seal failed	-	-	1	3.2	1
Wellhead failed	1	4.0	-	-	1
Unknown	1	4.0	3	9.7	4
Total	**25 (23)***	-	**31 (29)***	-	**56(52)***

* Figures in parentheses denote number of blowouts. For some blowouts two shallow gas handling "methods" are reported.

The *failed to stab kelly valve* incident stems from a low quality blowout description. The source only states that the well blew through the drillpipe. It is assumed that the operators were not able to stab the kelly valve.

Failed to close BOP occurred after first failing to operate the diverter.

BOP failed after closure occurred after the annular was closed and the jacket was evacuated due to a hurricane warning. After operators re-boarded the platform, they found the annular preventer to be leaking because the accumulator pressure was too low. The well was blowing saltwater, not gas.

BOP not in place was reported when a blowout occurred during running casing.

For at least 30 of the 52 shallow gas blowouts, the diverter system was in use, or was intended to be used. Operators may have intended to use the diverter system for some of the other blowouts, but these intentions were not mentioned. The use of diverter has three possible outcomes:

- *Diverted - no problem*
- *Failed to operate diverter*
- *Diverter failed after closure*

For one of the development wells, the drilling crew initially failed to operate the diverter due to a locked diverter overboard valve. When they managed to open the valve, the diverter line failed because of the sudden pressure increase when exposed to the gas; thus *failed to operate diverter* and *diverter failed after closure* were recorded in the SINTEF Offshore Blowout Database.

Diverters were used for 16 of the 23 development drilling shallow gas blowouts. The diverter system *failed to function as required* for five.

For 14 of the 29 shallow gas blowouts that occurred during exploration drilling, the diverter was used. It *failed to function as required* for nine.

Failed to operate diverter occurred four times. Three of these failures were related to diverter overboard valves, which were impossible to open. The fourth failure occurred because the diverter element was removed in order to run a hole opener.

Diverter failed after closure was reported 11 times. Eight of these were related to lines that were either worn or parted. The three remaining incidents were related to severe leakages in the main diverter packing element.

In the mid 1980s, diverter systems were subjected to heightened focus because they had been so unreliable in the past. The 1985 West Vanguard accident in the North Sea provoked this increased awareness of diverter systems [72]. Several studies were launched to improve diverter system design. Main recommendations from these studies were the diverter line diameter should be increased to reduce the gas velocity, and the number of bends and restrictions should be minimized in order to reduce the line erosion during a blowout situation [56].

The number of restrictions and the diameters of the lines are addressed in both Norwegian and U.S. regulations [3, 47].

From 1980-1992, the line diameters of diverter systems for the U.S. GoM OCS increased from 6 to 10-in. for jack-ups and platforms, and from 6 or 8-in. to 12-in. for semisubmersibles [44]. The number of bends was reduced to no more than two, and all right-angle turns were targeted.

The reliability of diverter systems has improved in previous years due to these modifications. The most recent shallow gas blowout in which the diverter system failed was in March 1989. Since then there have been eight shallow gas blowouts in which the diverter systems have been used. The diverter systems functioned as intended for all eight. Trend tests concluded that there is approximately an 85% probability of a reliability improvement.

Fracture at the casing shoe was listed seven times. For five of these blowouts (four exploration and one development) it was the result of closing in the well. One of the wells was closed in by a subsea shallow gas stack because the operators were not able to use the top side diverter. One *fracture at the casing shoe* occurred when drilling, and one occurred when diverting the well, leading to an underground blowout. This underground blowout stopped after ten days. Table 6.5 lists this blowout as a through annulus blowout, not an underground blowout.

SHALLOW GAS BLOWOUT CHARACTERISTICS

This section describes the experienced blowout characteristics that are relevant for risk evaluation.

Blowout flow paths and release points

Tables 6.5 and 6.6 lists the final flow paths vs. release points for shallow gas blowouts during development and exploration drilling. It is termed the final flow path because some blowouts may start with an initial flow path. A blowout may, for example, start flowing in the well bore annulus. If closing the well bore annulus, the well may flow outside the casing. Outside the casing will then be the final flow path. In most blowouts, the final flow path is the same as the initial flow path.

There are five different final flow paths; they are:

- *Through drillstring* (or tubing where relevant)
- *Through annulus* (the well bore annulus)
- *Through outer annulus* (between the casing strings)
- *Outside casing* (outside the outer casing or conductor)
- *Underground blowout* (subsurface blowout from one zone to another)

Table 6.5
Release Point vs. Final Flow Path for Development Drilling Shallow Gas Blowouts

Final flow path ⇒ Release point ⇓	Through drill-string	Through annulus	Through outer annulus	Outside casing	Un-known	Total
Diverted	-	9	1	-	-	10
Diverter syst.leak - line eroded	-	1	-	-	-	1
Diverter syst.leak - line parted	-	2	-	-	-	2
Drill floor - through rotary	-	3	-	-	1	4
Drill floor - top of drillstring	1	-	-	-	-	1
From platform wellhead	-	1	2	-	-	3
Subsea wellhead	-	-	1	-	-	1
Subsea - outside casing	-	-	-	1	-	1
Total	**1**	**16**	**4**	**1**	**1**	**23**

Table 6.6
Release Point vs. Final Flow Path for Exploration Drilling Shallow Gas Blowouts

Final flow path ⇒ Release point ⇓	Through drill-string	Through annulus	Through outer annulus	Outside casing	Un-known	Total
Diverted	-	6	-	-	-	6
Diverter syst.leak - line eroded	-	3	-	-	-	3
Diverter syst.leak - line parted	-	2	-	-	-	2
Drill floor - through rotary	-	1	-	-	-	1
From wellhead	-	-	1	-	-	1
Subsea crater	-	-	-	2	-	2
Subsea wellhead	-	6	1	-	-	7
Subsea - outside casing	-	-	-	6	-	6
Unknown	-	-	-	-	1	1
Total	**0**	**18**	**2**	**8**	**1**	**29**

Most shallow gas blowouts have a final flow path through the well bore annulus. The flow may be diverted without any problems, the diverter system may fail, or the flow may be released through the subsea wellhead.

The drillstring is seldom the final flow path for these types of blowouts. Only one blowout was reported in which the drillstring was the final flow path. One of the blowouts was listed as having a flow through drillstring as the initial flow path, but when the drillstring was closed the flow came through the annulus.

The final flow path for six of the incidents is reported as through the outer annulus (i.e., between casings).

Outside casing was reported for eight of the exploration drilling blowouts, and only one was reported for the development drilling blowouts. Blowouts that occur outside the casing are subsea releases. If the flow is large enough, it may result in cratering. Cratering is the worst scenario for bottom-supported installations in terms of loss of material assets and human lives. Cratering may cause the installation to sink or capsize. Two of the incidents involved cratering and resulted in both jack-up rigs sinking in the crater. No human lives were lost in these incidents.

Blowout flow medium

All blowouts, except one (shallow water), are reported with shallow gas as the flow medium. Six blowouts were reported with H_2S (two development and four exploration), and one development well was reported with some oil in the gas. Shallow gas blowouts do not cause severe *pollution.*

Blowout flowrates

The shallow gas blowout flowrates are in general unknown. However, they are usually thought to be high because of the large hole diameter, no fluid at all in the well, and short distance from influx to the surface.

Ignition of blowouts

Table 6.7 lists the experience related to shallow gas blowout ignition.

Table 6.7
Experienced Ignition for Shallow Gas Blowouts

PHASE	No ignition	Immediate ignition (< 5 min)	Delayed ignition 5 min - 1 hr	1 hr - 6 hr	6 hr - 24 hr	> 24 hr	Total
Development drilling	18	3	1	-	1	-	23
Exploration drilling	26	0	2	-	1	-	29
Total	**44**	**3**	**3**	-	**2**		**52**
	84.6%	**5.8%**	**5.8%**		**3.8%**	-	

As seen from Table 6.7, 85% of the shallow gas blowouts did not ignite. Three (6%) blowouts ignited immediately. Further, three ignited after 20 to 35 minutes. The remaining two ignited approximately 10 hours after the blowout started.

The overall trend regarding ignition of blowouts shows that the ignition frequency has decreased over the period 1980-1994 (see *Ignition Sources and Trends* on page 51). In light of the focus on reducing possible ignition sources, improving diverters systems, and improving procedures regarding hot work, this seems reasonable.

Blowout duration

Table 6.8 shows the duration of the shallow gas blowouts.

Table 6.8
Shallow Gas Drilling Blowout Duration

PHASE	Blowout duration ≤ 10 min	10 min - 40 min	40 min - 2 hr	2 hr - 12 hr	12 hr - 5 days	> 5 days	Unknown	Total
Dev. drlg	3	1	4	2	6	2	5	23
Expl. drlg	0	2	2	5	9	6	5	29
Total	**3**	**3**	**6**	**7**	**15**	**8**	**10**	**52**
	5.8%	**5.8%**	**11.5%**	**13.5%**	**28.8%**	**15.40%**	**19.2%**	

Method of well control

The primary means of regaining well control for 34 out of the 52 blowouts were bridging or depleting. Ten blowouts were controlled by pumping mud, two by squeezing cement, two by mechanical topside devices, and four were unknown.

Casualties

The two incidents in which the rigs sank did not cause any fatalities. Two of the blowouts that ignited involved fatalities. One was experienced during development drilling. The gas ignited immediately. This blowout caused the deaths of six persons. One blowout during exploration drilling ignited approximately 20 minutes after the gas release. One person died in this blowout.

Material losses

Due to lack of information, it is difficult to establish a detailed overview of the severity of the various blowouts. In Table 6.9, a coarse overview of blowout severity is given.

Table 6.9
Overview of Blowout Severity

Consequence class	Development drilling	Exploration drilling	Total
Small	19	25	44
Severe	3	1	4
Total loss	1	2	3
Unknown	-	1	5
Total	**23**	**29**	**52**

All shallow gas blowouts cause economic losses, even if no topside material damage is experienced. In the best case, only time is lost before the operation is back at the same stage as when the incident occurred. The total daily cost of a North Sea drilling rig exceeds (US) $100,000. A shallow gas blowout will always cause total time losses that exceed one day. Further, shallow gas blowouts frequently cause damage to topside equipment due to wear and tear caused by the gas flow. Further, the shallow gas well is frequently lost. On some occasions the wells also have to be killed and secured before they can be permanently abandoned. All these types of consequences are regarded as *small* in Table 6.9.

The four blowouts listed with *severe* consequences caused extensive topside damages, which required major repairs before the rigs could be put back into operation. Repairs for two development drilling shallow gas blowouts cost (US) $11 million and (US) $13 million. For the third development well shallow gas blowout, no repair costs were stated. For these three development well blowouts, the unknown amount of production losses will also add to the total losses. The exploration well shallow gas blowout was listed with a (US) $38 million repair cost (West Vanguard).

Two of the three blowouts reported as *total losses* were the two jack-up rigs that sank in the blowout crater. One of these blowouts was listed with (US) $32 million in losses. The third blowout was a production platform where the drilling rig and the living quarters were totally destroyed by a fire.

The worldwide shallow gas blowout experience since 1970 show that 12 shallow gas blowouts were reported with the *total loss* consequence (including the three mentioned above). Two of these were drill ships that sank because of reduced buoyancy. For one of these incidents, the ship would not have sunk if hatches near the sea level had been closed according to the procedures; the crew was not aware of the shallow gas danger and ignored the procedures. No detailed accident description was available in the other. It should be noted that no semisubmersible drilling rig has sunk as a result of a shallow gas blowout and reduced buoyancy.

The seven remaining *total loss* incidents were experienced on jack-up rigs. Four of the incidents were caused by seafloor cratering and subsequent capsizing. The three other incidents also caused the rig to capsize or to collapse into the water, but it is not known if the causes were cratering; two of the rigs were on fire.

DEEP DRILLING BLOWOUTS

All drilling blowouts *not* classified as shallow gas blowouts are classified as *"deep" blowouts*.

The main difference in blowout barriers (when drilling the deeper part of the well compared to the shallow part) is usually that two blowout barriers exist during "deep" drilling. The primary barrier is the drilling mud, and the secondary barriers are the mechanical devices designed for closing in the well annulus (a BOP) or the drillpipe (kelly valve or similar). When a mechanical barrier is activated during a kick situation, the well pressurizes. This requires that the formation fracture gradient be sufficiently high so that the pressure can be confined until the hydrostatic control is regained. If the formation fracture gradient is too low, an underground blowout or a blowout outside casing may result.

While a shallow gas kick frequently results in a blowout, a "deep" well kick usually does not result in a blowout.

Secondary barriers

A brief description of the secondary barriers during drilling is given below. Blowout barriers in general are also discussed in *Barriers in Well Operations* on page 17, along with several textbooks, among them [4].

During normal drilling, the secondary barriers are the blowout preventers (BOPs). The BOPs are located subsea for floating installations and topside for bottom-supported installations. BOPs are mainly used for closing in the well annulus, but most BOPs also include a device used for shearing the drillpipe and sealing the well. The annulus is usually sealed by closing an annular or a pipe ram preventer. The blind-shear ram preventer, which shears the pipe and seals the well, is regarded as an emergency device. Closing this preventer will significantly complicate the operation required to regain the hydrostatic control of the well.

If the well kicks through the drillpipe when the drillpipe is connected to the mud system, (i.e., not when the pipe is disconnected for tripping or adding an extra stand or joint) the pressure may be closed in by a valve located in the drillstring flow path. For drilling rigs with a rotary table, this will be a kelly valve. For drilling rigs with the more modern topdrives, a remote controlled valve inside the topdrive will close the drillstring flow path. If the drillpipe is disconnected when the well kicks, the kelly valve or the topdrive has to be stabbed in against the well flow to be able to stop the flow. If this does not succeed, the blind-shear ram preventer has to be activated.

Some drilling operators also use a check valve in the drillstring near the drill bit, which closes if the well flows through the drillstring. Several operators, however, decide not to install such a valve, because it may cause operational problems.

The casing, wellhead, and drillstring are also regarded as secondary barriers against blowouts. These barriers do not have to be activated in a kick situation.

One or more of the secondary barriers may not be available. This may be because the barrier itself failed (e.g., leakage in a wellhead connector), failed to activate (e.g., failed to close the BOP), or specific operations made the barriers unavailable (e.g., BOP nippled down to energize the casing seals). If the secondary barriers are unavailable and a kick occurs, the result will most probably be a blowout.

For other operations some of the barriers may be unavailable, (e.g., when running drill collars through the BOP) the blind-shear ram or pipe ram preventers cannot be used. When the drillpipe is out of the hole, the blind-shear ram is normally used to stop the well flow. Annular preventers may, however, also be used for this purpose, but only in an emergency.

The secondary barriers described above are the "normal" secondary barriers when drilling is in progress. During some specific operations different secondary barriers are used (i.e., when performing a production test on an exploration well or when running a wireline through the drillpipe).

After a kick is closed-in by the secondary barriers, the main goal is to re-establish hydrostatic control of the well. Several different methods exist to re-establish the hydrostatic control. The selection of method is related to the specific situation and the company's well control policy. The various methods applied, together with pros and cons, are described in several textbooks [4, 22].

DEEP DRILLING BLOWOUT EXPERIENCE

The experience presented in this chapter is based on the SINTEF Offshore Blowout Database (See *SINTEF Offshore Blowout Database* on page 33) for the period from January 1980 till January 1994 in the U.S. GoM OCS and the North Sea (UK and Norway). A total of 32 "deep" drilling blowouts has been recorded.

Table 6.10 lists the various installation types and well types.

Table 6.10
"Deep" Drilling Blowouts Experienced for Various Installations vs. Main Well Type

Installation type	Development drilling	Exploration drilling	Total
Jacket	5	-	5
Jack-up	6	9	15
Semisubmersible	-	8	8
Submersible	1	1	2
Total	**12**	**18**	**30**

As seen from Table 6.10, approximately the same number of "deep" drilling blowouts has occurred on jack-ups and semisubmersibles during exploration drilling. A comparison of blowout frequencies between these two types cannot be made because the number of wells drilled with the different types of installations is unknown.

For two blowouts (one exploration and one development), the type of installation listed is *submersible.* These drilling rigs are used in shallow water drilling; for these two blowouts, the water depth was 10 and 15 m (30 and 45 ft).

Jack-ups were used during six development well blowouts. At least four of these blowouts occurred when drilling wells for production platforms.

Figures 6.3 and 6.4 show the *number of wells to blowout* for "deep" blowouts and the corresponding regression lines and average lines.

Figure 6.3 indicates a slight reduction of the NWTB 1980-1994 for the "deep" blowout occurrences during exploration drilling. An overall significant trend in the exploration drilling "deep" blowouts NWTB could, however, not be concluded by any of the statistical trend tests used.

Figure 6.4 indicates a slight increase in the NWTB 1980-1994 for the "deep" blowout occurrences during development drilling. An overall significant trend in the development drilling "deep" blowouts NWTB could, however, not be concluded. For the development drilling "deep" blowout occurrences, many wells (approximately 2,500) had been drilled after the last blowout occurred. The regression analyses do not consider the wells drilled after the last blowout. These wells are so-called censored data; therefore, the average line lies above the regression line.

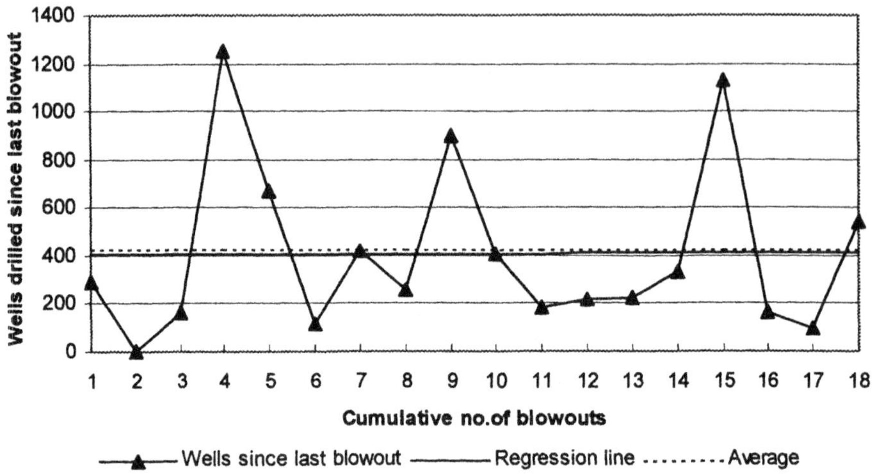

Figure 6.3 No. of wells to "deep" blowouts during exploration drilling, the associated regression line, and the average line

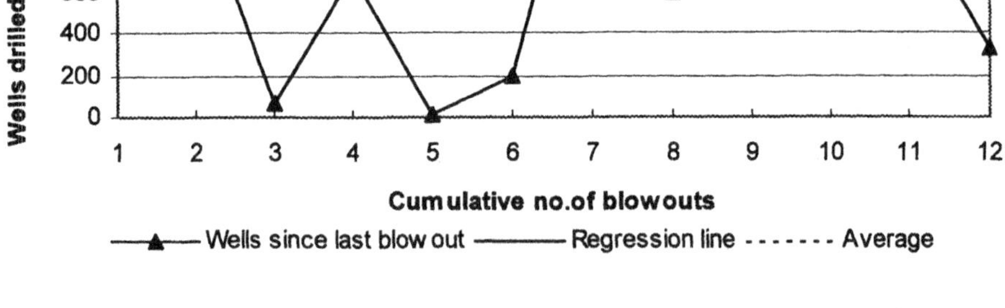

Figure 6.4 No. of wells to "deep" blowouts during development drilling, the associated regression line, and the average line

Table 6.11 presents the operations and activities in progress when the blowouts occurred for exploration and development drilling.

As can be seen from Table 6.11, *actual drilling, tripping out, and waiting on cement to harden* were the most frequent activities performed when a "deep" drilling blowout started. These activities represent 50% of all cases. Disregarding the blowouts with unknown activities, 67% of all "deep" drilling blowouts occurred during the following activities: *actual drilling, tripping out*, and *waiting on cement to harden.*

Table 6.11
Operations and Activities in Progress when "Deep" Drilling Blowouts Occurred

Operation ⇒	Drilling activity		Casing running		Well testing		Other/unknown operations		Total	
Activity ⇓	Dev.	Expl.	Dev.	Expl.	Dev.	Expl.	Dev.	Expl.	Dev.	Expl.
Actual drilling	3	6	-	-	-	-	-	-	3	6
Tripping out	1	2	-	-	-	1	-	-	1	3
Tripping in	-	-	1	-	-	-	-	-	1	-
Wait on cement	-	-	2	1	-	-	-	-	2	1
Nipple down BOP	-	-	1	-	-	-	-	-	1	-
Survey, wireline	1	-	-	-	-	-	-	-	1	-
Actual well test	-	-	-	-	1	-	-	-	1	-
Pull wireline	-	-	-	-	1	-	-	-	1	-
Gravel-packing	-	-	-	-	-	1	-	-	-	1
Pull casing	-	-	-	-	-	-	-	1	-	1
Maint. surface equipment	-	1	-	-	-	-	-	-	-	1
Unknown	2	2	-	-	-	-	-	4	2	6
Total	7	11	4 (3)*	1	2	2 (1)*	-	5	13(12)*	19(18)*

- Figures in parentheses denote number of blowouts. For some blowouts two activities are listed.

"DEEP" DRILLING BLOWOUT CAUSES

This section focuses on the causes of "deep" drilling blowouts. Since two barriers normally should be present while drilling, this section is divided into two main parts. The first part covers the causes of losing the primary barrier, mainly the hydrostatic control of the well. The second part covers the causes of losing the secondary barriers, mainly the topside barriers.

Loss of the primary barrier

When the primary barrier is lost, a well kick results. In terms of well control, it is important to detect the well kick as soon as possible in order to close in the well with a minimum influx. Small influxes are in general easier to handle than large influxes.

The ability to detect kicks has improved significantly since 1980. The control of the flow and pit volume has improved gradually during the period and is still improving. Today, compensating measures for rig heave are frequently used. Further, each mud pit has several sensors, and more sophisticated MWD (Measurement While Drilling) tools and computers for real time analysis of drilling data are used. The industry continuously focuses on improving kick detection.

According to drilling contractors, the properties of mud also have improved, and better procedures are used when changing mud densities.

One problem mentioned by the drilling contractors was that when drilling, the personnel sometimes believe too much in the well drilling plan and do not read the signals from the well.

Table 6.12 lists the experienced primary barrier failure causes for the kicks resulting in blowouts.

Table 6.12
Primary Barrier Failure Causes for "Deep" Drilling Blowouts

Primary barrier failure		Development drilling blowouts		Exploration drilling blowouts		Total
		No. of blowouts	Frequency (blowout/well)	No. of blowouts	Frequency (blowout/well)	
Too low hydro-static head	unexpected high well pressure	1	0.00010	2	0.00026	3
	too low mud weight	1	0.00010	3	0.00039	4
	swabbing	2	0.00020	2	0.00026	4
	improper fill up	-	-	1	0.00013	1
	gas cut mud	-	-	1	0.00013	1
	annular losses	-	-	1	0.00013	1
	while cement setting	2	0.00020	1	0.00013	3
	drilling into neighbor well	1	0.00010	-	-	1
	trapped gas	-	-	1	0.00013	1
	unknown why	3	0.00029	3	0.00039	6
Poor cement		-	-	1	0.00013	1
Well test string barrier failure		1	0.00010	-	-	1
Tubing plug failure		1	0.00010	-	-	1
Unknown		-	-	3	0.00039	3
Total		**12**	**0.00118**	**19(18)***	**0.00236****	**31(30)***

* Figures in parentheses denote number of blowouts. For some blowouts two primary barrier failures are reported.

** Based on number of blowouts

The relative distribution of causes of losing the primary barrier is rather similar to the shallow gas blowout causes shown in Table 6.3.

As seen from Table 6.12, "deep" drilling blowouts occur approximately twice as frequently during exploration drilling as during development drilling.

The main reason for losing the primary barrier during "deep" drilling is that the hydrostatic pressure becomes too low. The reasons for a too low hydrostatic head will be discussed in the following section. Otherwise, one "deep" drilling blowout was caused by poor cement, one by a barrier failure in the well test string, and one by a malfunctioning tubing plug.

The incident listed with *poor cement* was also listed with *too low hydrostatic head* as the cause for losing the primary barrier. During a cement squeeze job, gas propagated to a neighboring well. A poor cement job and a failed casing valve in the neighboring well caused a blowout through an intermediate or outer annulus.

The incident listed as a barrier failure in the test string occurred when preparing to reverse circulate after drill stem testing. The 3.5-in. drillpipe parted at 300 m (1,000 ft). The 275 bar (4,000 psi) well pressure blew off the 2-in. flowline to the bell nipple. The flowline fell down and ruptured a valve of the BOP control system, which caused lost accumulator pressure and disabled the BOP.

The tubing plug failure also occurred in association with well testing. The crew was testing a zone (not the planned production zone) prior to completing the well for production. The tail-pipe plug started to leak after pulling the wireline running tool. Then the wireline lubricator system failed to close in the well. The shear rams and the blind rams were then closed, but failed to stop the flow.

Swabbing

Swabbing is a frequent cause of losing the hydrostatic barrier, and hence leads to "deep" drilling blowouts. However, swabbing is a more dominant cause for shallow gas blowouts.

Three of these incidents occurred in association with tripping out of the well with the string. The fourth, however, occurred when pulling a directional wireline survey tool.

The discussion related to swabbing in *Shallow Gas Blowout Causes* on page 65 is also relevant for "deep" drilling blowouts.

Unexpected high well pressure/too low mud weight

One of the three *too low mud weight* incidents was caused by mud which was mixed to a density different from the density specified in the well plan (exploration well). Another incident most likely occurred when lowering the casing string into a development well. The remaining incident occurred while actually drilling. It is likely that the incident was caused by an *unexpected high well pressure.*

The four blowouts listed with *unexpected high well pressure* occurred while drilling ahead.

As anticipated, *unexpected high well pressure* is reported more frequently as the cause for losing the primary barrier during exploration drilling blowouts than for development drilling blowouts. This is the same result observed for the shallow gas blowouts. The main cause of this difference may be explained by lack of reservoir knowledge.

The occurrence of blowouts with *unexpected high well pressure/too low mud weight* as the cause for losing the primary barrier does *not* seem to reveal any trend during 1980-1994.

Improper fill up

Improper fill up was reported once as the cause of losing the primary barrier. The drilling crew had pulled 25 stands of drillpipe without filling the hole. The scheduled filling was every fifth stand. This cause was not observed for the shallow gas blowouts, but because of a limited description of the incident, it is difficult to distinguish this cause from swabbing.

While cement setting

Three blowouts occurred some time after the cement was in place. The BOP was nippled down for all three wells. In two of the blowouts, gas started to migrate through the 10 3/4 x 16-in. casing annulus. The description of the third incident did not include such details.

It should be noted that these three blowouts occurred in 1988, 1989, and 1991, which is when so-called "gas tight cement" should have been available. It was not stated what type of cement was used.

Other

Annular losses as causes of the well kick were only reported once for "deep" drilling blowouts.

Gas cut mud is reported as the cause of losing the primary barrier for one exploration well blowout. First, the crew experienced a drilling break and checked for flow. Then the well was circulated twice before drilling was resumed at a controlled rate, which is when the well began to flow.

Drilling into the neighboring well is reported once as the cause for losing the primary barrier. The incident occurred in December 1985. The crew drilled into a gas-lifted well. Little information exists about this incident. The database contains five similar incidents - three in the U.S. GoM in the 1970's, one in Dubai in 1982, and one in Trinidad in 1991.

Trapped gas reportedly caused the well kick, resulting in a "deep" drilling blowout. The crew was abandoning the well and had plugged the well with a cement plug. First, they perforated the casing and observed pressure for a build-up, then they cut the casing and checked for flow. When they started to pull the casing, pressurized gas trapped behind the casing blew the riser clean of mud. Another "deep" drilling blowout included trapped gas below the BOP. That was, however, not the cause of the kick, but a result of improper kick circulation.

Three development and three exploration well blowouts were listed with ***unknown*** causes of losing the hydrostatic head.

Loss of the secondary barrier

Table 6.13 lists the causes of losing the secondary barrier.

Table 6.13
Secondary Barrier Failures for "Deep" Drilling Blowouts

Secondary barrier failures	Development drilling		Exploration drilling		Total
	No. of blowouts	Distribution (%)	No. of blowouts	Distribution (%)	
String safety valve not available	1	8.3	2	10.0	3
Failed to stab kelly valve	-	-	2	10.0	2
Failed to close BOP	3	25.0	2	10.0	5
BOP failed after closure	2	16.7	3	15.0	5
BOP not in place	2	16.7	1	5.0	3
Fracture at casing shoe	1	8.3	1	5.0	2
Poor cement	-	-	1	5.0	1
Casing valve failure	-	-	1	5.0	1
Wireline BOP/lubricator not installed	1	8.3	-	-	1
Casing leakage	1	8.3	3	15.0	4
Unknown	1	8.3	4	20.0	5
Total	**12**	-	**20(18)***	-	**32(30)***

* Figures in parentheses denote number of blowouts. For some blowouts two secondary barrier failures are reported.

Drillstring safety valve not available

Three blowouts were listed with *string safety valve not available* as the failed secondary barrier. For the development drilling blowout this was assumed to be the cause because, while lowering scraper and mule shoe, the well blew through the drill stem. No attempt to close a drillstring safety valve was mentioned. Further, it should be noted that there were no shear rams in the BOP stack. The well bridged after three hours.

For one of the exploration drilling blowouts, the kelly valve was 3.7 m (12 ft) in the air and could not be reached when the well flowed in. The stand-pipe valve could not be closed due to high differential pressure. Also, it seemed that the BOP stack did not include shear rams, because the use of shear rams was not discussed in the blowout investigation report.

For the other exploration well blowout, the drillstring safety valve could not be closed because coiled tubing was running through the valve. The blind-shear ram was closed to control the surface flow.

Failed to stab kelly valve

Two blowouts were listed with *failed to stab kelly valve* as the cause for losing the secondary barrier. In both incidents, stabbing the valve against the flow was impossible. One of these two stabbing attempts was with a top drive. After failing to stab, the crew unscrewed one stand at the drill floor and attempted to stab the kelly valve, which also failed. The blind-shear rams were closed to control the surface flow for both incidents.

BOP failures

For 13 of the 32 blowouts the well kick developed to a blowout because of BOP failures or mal-operation of the BOP. One blowout was reported with both *failed to close the BOP* and *BOP failed after closure*.

Failed to close the BOP

Five blowouts were reported with *failed to close the BOP* as the secondary barrier failure. For one development and one exploration well blowout, there were no failures in the BOP, but the BOP was closed late. One was closed late due to inattention to the well situation. For the second, trapped gas shallow in the well came in, and there was not enough time to close the BOP before the gas was released. There was a blowout on a jacket, and the data source only stated that the crew failed to close the BOP. For the fourth *failed to close BOP* incident, the BOP control system was disabled because the well test flowline fell on a control system valve, which resulted in a loss of accumulator pressure. This incident was during development drilling.

The fifth *failed to close BOP* incident also reported that the *BOP failed after closure* as the cause for losing the secondary barrier. First, the subsea BOP middle pipe ram preventer was closed. This preventer failed to seal the well off; the source stated that it was probably due to scores in the drillpipe. The gas flowing in the riser ignited immediately upon reaching the rig. The explosion severed the hydraulic lines, therefore closing the shear ram preventer was impossible.

BOP failed after closure

After running a bridge plug in the tail-pipe in connection with a development well test, the plug started to leak or would not stay in place. The wireline lubricator started to leak after the crew pulled the tools. The crew closed the shear ram preventer, but the flow started again. Then they closed the blind ram preventer. The gas then started to flow via the BOP control system lines.

During circulation of a kick in an exploration well, the flexible choke line on the subsea BOP parted. The subsea BOP choke valves did not seal off the flow because the valves were severely eroded.

The annular preventer was closed to circulate out a kick in another exploration well. The pressure below the annular preventer was approximately 300 bar (4,350 psi) when the pipe started to strip out of the well. The crew attempted to close the blind ram preventer, but some pipe remained in the well. The stack was not equipped with a shear ram preventer. The BOP also included a variable and a fixed ram preventer. No attempts were made to close these ram preventers.

Another *BOP failed after closure* is from a low quality source. During circulation of a kick, an explosion occurred near the shale shaker. To stop the flow, the shear ram preventer was closed. It seems likely that other means, such as choke/kill valves or a pipe ram preventer, should have been used. It is therefore assumed that the BOP was malfunctioning.

Ocean Odyssey blowout, UK sector of the North Sea, 1988.
The semisubmersible was drilling at 4,928 m when the high pressure well kicked. During the kick circulation, first the choke line then the choke valves failed. The released gas ignited when it reached the surface. One person was killed.

BOP not in place

The three incidents involving *BOPs not in place* were all similar. The BOP had been nippled down after cementing the casing.

BOP performance in general

For some U.S. GoM OCS blowouts, a blind-shear ram preventer was not included in the BOP stack. If a blind-shear ram preventer had been used, the flow might have been stopped earlier (i.e., the blowout would have been less serious). It is not mandatory in U.S. OCS waters to use blind-shear rams for surface BOPs, only in subsea BOPs [15].

BOP reliability is important in kick control. Usually the BOP functions as intended when a kick occurs, and the kick may therefore be circulated out.

Quantifying the BOP reliability, or trends in BOP reliability, is impossible based only on the blowout data. However, BOP reliability has been investigated through numerous studies at SINTEF. Holand sums up the results related to subsea BOP reliability in [32, 33]. BOP failures from 250 exploration wells in the Norwegian sector of the North Sea have been collected and analyzed. In addition, the reliability of surface BOPs has been investigated in a separate study [31]. This study is based on BOP failures from 50 development wells.

Some of the conclusions from these studies are:

- The subsea BOP failure rate has shown a decreasing tendency in the beginning of the 1980s and seems to have stabilized at a certain level since 1984/85.
- Currently, critical failures in the preventers, connectors and choke/kill valves are rare in North Sea subsea BOPs.
- Total malfunction of a *subsea BOP hydraulic control system* has not been observed. Losing one of the redundant pods is, however, fairly common.

- Subsea BOP failures are costly, and approximately 3% of the total rig time required to drill a well is lost because of BOP failures and the associated repair activities.
- Critical failures in surface BOPs occur far more often than in subsea BOPs. Failures affecting both the main preventer sealing and loss of BOP control have been observed.
- Surface BOP failures only cause 0.36% of the total drilling rig time to be lost because of BOP repair.

West Hou Inc. compared the rig downtime caused by subsea BOP stacks in the U.S. GoM OCS and UK with results from Holand [32]. They concluded that the average downtime caused by BOP failures is nearly identical in Norway, UK, and U.S. GoM; however, the equipment failures that contribute to this downtime vary significantly [12].

The equipment used in the U.S. GoM OCS is fairly identical to North Sea equipment. One difference is that U.S. GoM OCS subsea BOP stacks do not use acoustic back-up control systems. U.S. companies only apply these systems in Alaska. Further, in U.S. GoM OCS surface BOPs it is not mandatory to install blind-shear ram preventers [15].

The *BOP technology* today is similar to the technology in the beginning of the 1980s. Most of the introduced changes are caused by the need for equipment that may be used for higher pressures and temperatures. Otherwise, some equipment has been changed because of experienced operational problems. New equipment includes new bonnet seals for the Cameron UII preventers and improved designs of choke and kill valves, which in the beginning of the 1980s, frequently failed to seal high pressure [32]. In addition, variable bore rams have become more common and may close around different sizes of tubulars; thus, on average, more ram preventers will be available for closing in a well when drillstring components of various diameters are run in the well. Hydril recently launched a new ram type BOP, a so-called Compact BOP. A new bonnet seal design permits low torque make up of the bonnets. This bonnet seal principle allows more elastic strain of the BOP body; therefore, the overall ram BOP weight is reduced by approximately 20% [36].

Drilling contractors and BOP suppliers stated that, in general, the quality of North Sea *BOP maintenance* has improved. Occasionally, when the rig rates are low, non-original equipment manufacturers do the BOP maintenance, which is likely to reduce the quality of maintenance.

Testing of BOP systems in the U.S. GoM OCS was almost identical to the testing performed in Norway before the Norwegian testing regulations were changed in 1991. The main difference was that in the U.S. it was mandatory to perform low pressure tests. Norwegian regulations in 1991 prolonged the test interval for high pressure BOP tests from one to two weeks. These BOP pressure tests also include low pressure tests. Biweekly BOP function tests replace the pressure tests. A pressure test of the choke and kill lines is also included in these biweekly BOP function tests. Holand concluded that these changes of BOP testing would not significantly affect the *subsea BOP* safety availability [32]. However, for *surface BOPs,* it was concluded that these changes in BOP test frequency would reduce the safety availability [31]. These conclusions were based on an evaluation of the experienced BOP failures and how they were detected.

Fracture at casing shoe

Fracture at casing shoe occurred twice, once on an exploration well and once on a development well. The development well blowout occurred when the crew was circulating out a kick. The formation broke down at the casing shoe and the gas channeled to the neighboring well, where it came to surface.

The second *fracture at casing shoe* occurred during a kick situation. The drillpipe was stuck with 7 bar (100 psi) pressure on the drillpipe and 43 bar (620 psi) on the casing when the crew noted two gas boils, one located around the pipe and the other outside the pad of the jack-up.

Poor cement and casing valve failure

Poor cement and *casing valve failure* were reported on the same blowout. A cement squeeze job on one well failed, allowing gas to propagate to another well. A poor cement job and a failed casing valve caused a blowout through an intermediate or outer annulus on the other well. The underground blowout bridged after four days. The casing valve was repaired and closed.

Wireline BOP/lubricator not installed

A *wireline BOP/lubricator not installed* incident occurred when the wireline directional survey tool was being pulled from the drillpipe. A rapid flow was observed through the drillpipe, and workers cut the wireline, leaving 1,500 m (4,900 ft) of wireline inside the drillpipe, and then closing a drillstring safety valve.

Casing leakage

Casing leakage was reported as the secondary barrier failure for four blowouts: one development well and three exploration wells.

For two of the exploration wells, the casing ruptured just below the BOP during circulation of a kick. Both of these blowouts bridged. The third exploration well with *casing leakage* had a hole in the casing which resulted in seafloor bubbles.

The *casing leakage* incident on the development well occurred during stripping drillpipe out of the well through the annular preventer after a well kick. The drillpipe elevators failed and the drillpipe fell down. The formation at the intermediate shoe fractured and the surface pressure increased, causing the casing to rupture at 996 m (3,268 ft). This fracture resulted in an underground blowout. The well finally bridged after several months.

Unknown

Five incidents are reported with *unknown* causes for losing the secondary barrier.

"DEEP" DRILLING BLOWOUT CHARACTERISTICS

This section describes the "deep" drilling blowout characteristics that are relevant for risk evaluation.

Blowout flow paths and release points

Tables 6.14 and 6.15 list the final flow path vs. the release point for "deep" blowouts during development and exploration drilling, respectively.

Table 6.14
Release Point vs. Final Flow Path for "Deep" Development Drilling Blowouts

Final flow path ⇒ Release point ⇓	Through drill-string	Through annulus	Through outer annulus	Outside casing	Under-ground blowout	Un-known	Total
Drill floor - choke manifold	-	1	-	-	-	-	1
Drill floor - through rotary	-	3	-	-	-	-	3
Drill floor - top of drillstring	2	-	-	-	-	-	2
From wellhead	-	1	2	-	-	-	3
No surface flow	-	-	-	-	1	-	1
Unknown	-	1	-	-	-	1	2
Total	**2**	**6**	**2**	**-**	**1**	**1**	**12**

Through annulus is the most common final flow path during "deep" development drilling blowouts. Fifty percent of the blowouts were flowing through the annulus.

Outside casing was not reported for any of the development drilling blowouts. This final flow path was rare for development drilling shallow gas blowouts as well (see Table 6.5). Blowouts outside the casing cause subsea releases. None of the development well "deep" blowouts caused subsea releases.

Table 6.15
Release Point vs. Final Flow Path for "Deep" Exploration Drilling Blowouts

Final flow path ⇒ Release point ⇓	Through drill-string	Through annulus	Through outer annulus	Outside casing	Under-ground blowout	Un-known	Total
Drill floor - through rotary	-	2	-	-	-	-	2
Drill floor - top of drillstring	3	-	-	-	-	-	3
From wellhead	-	1	2	-	-	-	3
No surface flow	-	-	-	-	2	-	2
Shaker room	-	1	-	-	-	-	1
Subsea BOP (BOP choke line)	-	1	-	-	-	-	1
Subsea - outside casing	-	-	-	3	-	-	3
Unknown	-	-	-	-	-	3	3
Total	**3**	**5**	**2**	**3**	**2**	**3**	**18**

Four of the 18 exploration drilling blowouts caused subsea releases. In two of these incidents, the flow was limited. For the two others, the flow was significant, and one ignited when the gas reached the surface. In general, subsea releases are far more frequent in exploration well blowouts than in development well blowouts. This frequency was also observed for the shallow gas blowouts (Tables 6.5 and 6.6).

Only about 30% of the exploration well blowouts had final flow paths through drilling annuluses.

Blowout flow medium

Table 6.16 shows an overview of the blowout flow mediums for "deep" drilling blowouts.

Blowouts, which are recorded both with gas and oil as flow mediums, have been listed as oil blowouts, and blowouts with both gas and condensate have been listed as condensate blowouts.

Table 6.16
Blowout Flow Mediums for North Sea and U.S. GoM OCS "Deep" Drilling Blowouts

PHASE	Blowout flow medium				Total
	Gas	Cond-ensate	Oil	Un-known	
Development drilling	10 *	1	1	-	12
Exploration drilling	15**	-	2	1	18
Total	**25**	**1**	**3**	**1**	**30**

* One was gas-lift gas from the neighbor well.
* One was gas that was trapped behind the casing.

As seen from Table 6.16, most of the "deep" drilling blowouts comprise gas and only a few flow oil or condensate. This does not mean that drilling for gas is significantly more risky than drilling for oil. Several of the gas blowouts were experienced when drilling "oil wells". Many oil reservoirs have a cap of free gas above the oil zone. Gas also may have accumulated in minor pockets above the main reservoir. This means that during drilling the gas zone will be reached before the oil zone. The well may then kick if the hydrostatic pressure is inadequate because the gas zone will have approximately the same pressure as the oil zone below. The coning effect may or may not bring some oil in the flow.

Even if the oil zone is penetrated when the blowout starts (i.e. blowout caused by swabbing), the flow on the surface will, if measured by volume, normally mainly consist of gas. If, however, the reservoir has no gas cap, or drilling beside the gas cap is possible, a well kick will usually be easier to handle because of the reduced expansion with decreasing pressure. Then oil will be a major part of the flow medium. Further, if an oil well is cased down to the oil zone and a blowout occurs, oil will be a major part of the flow medium.

Blowout flowrates

Flowrates of the blowouts that were experienced are usually not reported. For some blowouts, flowrate values exist, but for most blowouts they do not exist.

Flowrates are important figures in risk and environmental analyses. To establish a realistic distribution of expected flowrates for specific fields, field specific productivity data should be compared to blowout experience with respect to remaining restrictions in the wells during the blowout situations. For several blowouts there are significant flow restrictions that will reduce the flow.

Blowout pollution

None of the "deep" drilling blowouts in the North Sea and the U.S. GoM OCS after January 1980 caused severe pollution. Two incidents involved minor pollution. The most significant pollution of these two incidents was reported to be some few cubic meters of oil in the sea.

However, of 110 worldwide "deep" drilling blowouts after January 1970, six blowouts reportedly caused major spills. Five of these incidents occurred during exploration drilling and one during development drilling.

Severe pollution incidents may result from a blowout, but U.S. GoM OCS and North Sea experience show that blowouts seldom cause severe pollution.

Ignition of blowouts

Table 6.17 lists the experience related to "deep" drilling blowout ignition.

Table 6.17
Experienced Ignition for "Deep" Drilling Blowouts

PHASE	No ignition	Immediate ignition (< 5 min)	Delayed ignition 5 min - 1 hr	1 hr - 6 hr	6 hr - 24 hr	> 24 hr	Total
Development drilling	11	1	-	-	0	-	12
Exploration drilling	14	1	-	-	3	-	18
Total	**25**	**2**	**-**	**-**	**3**	**-**	**30**
	83.3%	**6.7%**			**10.0%**		

As seen from Table 6.17, 83% of the "deep" drilling blowouts did not ignite. Two (7%) blowouts ignited immediately. The remaining three (10%) ignited seven, eight, and twelve hours after the blowouts started. Ignition sources and trends are discussed in *Ignition Sources and Trends* on page 51.

Blowout duration

Table 6.18 presents the blowout durations for "deep" drilling blowouts.

Table 6.18
"Deep" Drilling Blowout Duration

PHASE	Blowout duration ≤ 10 min	10 min - 40 min	40 min - 2 hr	2 hr - 12 hr	12 hr - 5 days	> 5 days	Unknown	Total
Dev. Drlg	1	-	-	3	4	3	1	12
Expl. Drlg	1	1	1	2	7	2	4	18
Total	**2**	**1**	**1**	**5**	**11**	**5**	**5**	**30**
	6.7%	**3.3%**	**3.3%**	**16.7%**	**36.6%**	**16.7%**	**16.7%**	

Method of well control

Bridging was the primary "method" for regaining well control for 13 out of the 32 blowouts. Control was regained for four blowouts by pumping mud, seven were controlled by using BOPs, one by other mechanical topside devices, and the methods of control are unknown for seven "deep" drilling blowouts.

Casualties

Four of the blowouts involved casualties. In total, 11 persons died. Two incidents caused one fatality each, four died in one incident, and five died in another. All these blowouts occurred during exploration drilling, and all blowouts ignited. So all the exploration well blowouts that ignited involved fatalities. It should be noted that for three of the four incidents, the blowouts ignited several hours after the first release of gas (Table 6.17). The last incident, which caused the death of one person, ignited immediately.

Material losses

It is difficult, due to lack of information, to establish a detailed overview of the material losses from the various blowouts. Table 6.19 shows an overview of the "deep" drilling blowout severity. Note that the severity is related to platform or rig damage, not the cost of the well control operations.

Table 6.19
"Deep" Drilling Blowout Severity

Consequence class	Development drilling	Exploration drilling	Total
Small	9	12	23
Severe	-	2	2
Total loss	-	1	1
Unknown	3	3	6
Total	**12**	**18**	**32**

All blowouts cause economic losses, even if no topside material is damaged. At best, only time is lost before the operation is back at the same stage as it was when the incident occurred. The well is, however, often lost. Killing operations may last for months before the well is completely secured.

In Table 6.19, *only topside damages are considered* in the severity listing. The cost of well control efforts is not evaluated and therefore not included. The well-known Treasure Saga blowout in the North Sea in 1989, which required more than 200 days of well control activities, is included in the *small* consequence class. The total cost of this blowout was nearly (US) $300 million.

The two blowouts listed with *severe* consequences caused extensive topside damages, which required major repairs before the rigs could be put back in operation. One company reported (US) $15 million in repair costs; the other company's blowout repair costs are unknown. However, the latter company did rebuild the rig. Both of these blowouts involved fatalities.

The rig which experienced the blowout listed with *total loss* as consequence was rebuilt and put into service two years later. Repair costs were (US) $30 million. This incident occurred in 1980. After 1980, no *total loss* incidents have occurred during "deep" well drilling. From 1970-1979, three *total loss* incidents in the North Sea and U.S. GoM OCS occurred.

Of 110 "deep" drilling blowouts worldwide after January 1970, 16 blowouts were listed with consequence class *total loss,* and ten were reported as *severe*.

CHAPTER SEVEN

Completion Blowouts

Completion blowouts occur during well completion activities. Well completion activities involve installing equipment or undertaking operations required to produce a well after the drilling is completed. This usually includes preparation for and running of the production tubing, and installation of the Christmas tree. If the wells, for instance, are gravel-packed, or are in any other ways prepared before running the tubing, this is regarded as a part of the completion activities.

The complexity of a well completion varies significantly; some are simple, while others are complex. The complexity will vary from field to field and from operator to operator. Complexity is mainly dependent on the reservoir, the oil company's preferences and requirements, and the government requirements.

The complexity depends on whether there are:

- gravel-pack, sand screens or nothing
- dual or single completions
- artificial lift (now or later)
- non-corrosive equipment
- equipment for downhole chemical injection
- dual/single downhole safety valve
- annulus safety valve
- etc.

In this book no distinctions have been made regarding the equipment included in the various well completions. This is because the information required to make such distinctions is not available and the total number of completion blowouts is low.

The SINTEF Offshore Blowout Database includes seven completion blowouts from January 1980 - January 1994 in U.S. GoM OCS and the North Sea (Norwegian, UK).

TRENDS IN COMPLETION BLOWOUT FREQUENCY

Of the seven completion blowouts, *one* occurred in 1980, *five* in 1981, and *one* in 1987. No specific reason explains the accumulation of completion blowouts in 1981. The number of wells completed was not significantly different from other years. The five blowouts in 1981 occurred on different fields, and the events were spread from January until the end of October that year.

This indicates that there is a trend in the blowout probability, and the present blowout probability level is lower than the average blowout frequency would suggest. The trend in completion blowouts was investigated and was found to be significant. The results of the trend analyses are discussed in *Wireline Blowouts* on page 138. Significant overall trends in the blowout frequencies for *exploration drilling, development drilling*, and *workovers* could not be identified (Chapters 5, 6 and 8).

Comments related to Tables 7.2 and 7.3 focus on the problems related to establishing the secondary barriers when the well kicked. The main problem for several blowouts was the BOP stacks did not include a blind-shear ram preventer, which might have prevented the blowout. It is not mandatory in U.S. OCS waters to use blind-shear rams in surface BOPs, only in subsea BOPs [15].

COMPLETION BLOWOUT CAUSES

Table 7.1 shows the operations and activities in progress when the completion blowouts occurred.

Table 7.1
Operations and Activities when Completion Blowouts Occurred

Operation ⇒ Activity ⇓	Running well equipment	Well testing incl. preparations	Circulating	Unknown	Total
Tripping out	-	1	-	1	2
Actual circulation	-	-	1	-	1
Killing	-	-	1	-	1
Perforating	-	1	-	-	1
Gravel-pack	1	-	-	-	1
Unknown	-	-	-	2	2
Total	**1**	**2 (1)***	**2**	**3**	**8 (7)***

* Figures in parentheses denote number of blowouts. For some blowouts, two activities are listed.

Since there are few occurrences of completion blowouts, specific trends in operations and activities when the incidents occurred cannot be identified. The only thing to note is that two of the seven completion blowouts occurred when tripping out of the well.

Table 7.2 lists the causes for losing the primary and secondary barriers.

Table 7.2
Primary and Secondary Barriers Failures for Completion Blowouts

Loss of primary barrier⇒	Too low hydrostatic head			Tubing to annulus leakage	Unknown	Total
Loss of secondary barriers ⇓	Annular losses	Swabbing	Too low mud weight			
String safety valve failed	-	-	1	-	-	1
String safety valve not available	-	-	1	-	-	1
Failed to stab kelly valve	-	1	-	-	-	1
Insufficient frictional back pressure	-	1	-	-	-	1
Failed to close BOP	1	-	-	1	-	2
Unknown	-	-	-	-	1	1
Total	**1**	**2**	**2**	**1**	**1**	**7**

In the incident listed with *too low mud weight* and *string safety valve failed,* the well started to flow when the drilling crew reduced the mud weight from 1,680-1,200 kg/m^3 (14 to 10 lb./gal). The kelly valve would not close. The kelly valve and the stand-pipe valve leaked. The crew also attempted to pump into the well, but failed. It seems that the BOP did not include a blind-shear ram preventer, because the well was killed by installing a shear ram preventer in the BOP stack.

The incident listed with *too low mud weight* and *string safety valve not available* was poorly described. The well started to flow because the well was not properly killed, and a so-called "back pressure valve" had been removed before starting the completion operations.

For the incident listed with *swabbing* and *failed to stab kelly valve,* the source only states that the incident occurred while pulling out of the hole, and the blowout was controlled by freezing the pipe with dry ice after one day. It is assumed that the drilling crew attempted to stab a valve, but failed. It is likely that the BOP stack did not include a blind-shear ram preventer.

For the incident listed with *swabbing* and *insufficient frictional back pressure,* the crew had just perforated the upper sand and circulated the annulus before they started to pull out of the well. The well then started to flow through the open pipe. The well was then closed in and the crew attempted to strip into the well through the annular. The 3.5-in. pipe, the slips, and the rotary bushing were then blown out.

For the incident listed with *annular losses* and *failed to close BOP,* the mud was lost to a lower productive zone. The mud weight was reduced and periodically filled up. The gravel-pack equipment was being run when the upper productive zone kicked. The pipe rams did not have the correct dimension, and the stack did not include a blind-shear ram preventer or an annular preventer. The blind ram was closed on 4.5-in. pipe.

For the incident listed with *tubing to annulus leakage* and *failed to close BOP,* coiled tubing was used to start the flow in a newly perforated zone. The well was then shut in. Thereafter, the crew started pulling on the tubing in order to open the bypass valve to circulate and kill the well. The 2 7/8-in. tubing then jumped threads on a 2 7/8-in. coupling just above the BOP and below the control head, and thereby parted. The tubing fell down inside the BOP before the lower and upper pipe ram preventers were closed. The crew experienced problems in closing the annular and the blind ram preventer. The well was flowing through the open tubing. The BOP did not have a blind-shear ram preventer, and there were no downhole barriers.

The source information for the last completion blowout listed with *unknown* (regarding the primary and the secondary barrier) was of very poor quality. The sources said only that it was a gas blowout because a high pressure joint in the line broke loose, and it occurred during completion activities.

COMPLETION BLOWOUT CHARACTERISTICS

Blowout flow paths and release point

Table 7.3 lists the final flow path vs. the release point for blowouts during completion activities.

Table 7.3
Release Point vs. Final Flow Path for Completion Blowouts

Final flow path ⇒ **Release point ⇓**	**Through drillstring**	**Through tubing**	**Through annulus**	**Total**
Drill floor - drillpipe valve	1	-	-	1
Drill floor - through rotary	-	-	2	2
Drill floor - top of drillstring	2	-	-	2
Drill floor - top of tubing	-	1	-	1
Unknown	-	1	-	1
Total	**3**	**2**	**2**	**7**

Most blowouts that occur during completions result in flow through the tubing or the drillstring/work string. For several of these blowouts, the BOP stack did not include a blind-shear ram. If a blind-shear ram had been employed, these blowouts could have been stopped at an earlier stage and, in many cases, would not have been categorized as blowouts. It is not mandatory to use blind-shear rams in U.S. OCS surface BOPs [15].

Two of the completion blowouts flowed through the annulus. During one of the blowouts, the drillpipe was blown out of the hole. During the other, the pipe rams did not fit the pipe inside the BOP, and the BOP stack did not include an annular and a blind-shear ram preventer.

Blowout flow medium

Six of the seven blowouts were listed with *gas* as the flow medium, while the last blowout was listed with both *oil* and *gas.*

Blowout pollution

None of these seven blowouts caused severe pollution. Five were reported with no pollution, and two were listed with light sheen as pollution.

However, of 18 worldwide completion blowouts since January 1970, two caused large pollution, one off Iran in 1971 and one off Nigeria in 1980.

Ignition of blowouts

One of the seven completion blowouts ignited twelve hours after the blowout started and severely damaged the platform. The other six blowouts did not ignite.

Blowout duration

Table 7.4 presents the durations of the completion blowouts.

Table 7.4
Completion Blowout Duration

PHASE	Blowout duration							
	≤ 10 min	10 min - 40 min	40 min - 2 hr	2 hr - 12 hr	12 hr - 5 days	> 5 days	Un-known	Total
Completion	1 14.3%	-	-	-	4 57.1%	2 28.6%	-	7

Casualties

None of the completion blowouts involved casualties.

Material losses

The only blowout causing severe topside damage was the blowout that ignited. Otherwise, only minor topside damages were reported.

CHAPTER EIGHT

Workover Blowouts

Although most offshore blowouts occur during drilling, more deep well blowouts occur during workovers than during development drilling (Table 5.1). Only one of the 19 workover blowouts occurred in the North Sea, while the remaining occurred in the U.S. GoM OCS. The U.S. GoM OCS is a more mature area than the North Sea. In the U.S. GoM OCS, the number of wells in production is at approximately the same level as it was in 1980; whereas, in the North Sea (UK and Norway), there were approximately 2.5 times as many wells in production in 1993 as in 1980. Because of the increased number of wells and the average increased ages of the wells, the required number of workovers to be performed in the North Sea is increasing from year to year.

Workover blowouts are more likely to cause severe pollution than drilling blowouts because the wells are cased down to the productive zone when the blowouts occur. If the well blows out, the content of the flow seen topside is dependent on whether the well is perforated in an oil, condensate, or gas zone. Drilling blowouts will mainly be gas blowouts. This is confirmed in Table 5.6.

A well workover as defined in this book is a well overhaul/repair operation that normally involves complete or partial pulling of the production tubing. In recent years, other methods for performing well overhaul/repair have become more common, particularly snubbing and coiled tubing operations. Coiled tubing operations are expected to be used more and more in the future. Bedford describes a very positive experience with using coiled tubing in well maintenance [10]. The use of coiled tubing and snubbing will reduce the need for conventional workovers. The motivation for these methods is primarily to reduce costs. Whether these operations cause increased or reduced blowout risk has not been verified.

The blowout barriers present during workover operations are normally the same barriers that exist for development drilling. However, it is important to note that there are several differences [55]:

- During workovers, a productive zone is exposed nearly all the time (i.e., a flow is possible). For drilling, a productive zone is exposed only for a short duration of the total drilling period.
- Solids-free workover fluids are usually used during workovers. A mud filter cake, which during drilling acts as a seal against the formation, will not be created. This means that during workovers there are normally continuous losses to the formation.
- In a workover, the well can be closed in with higher pressures than during drilling because formation breakdowns on shallow casing shoes are less likely to occur.
- Bullheading is a kill method that has a high success probability for workover kicks, compared to drilling kicks.
- In workovers, there is less knowledge about the casing condition, because the casing strings have been in the well for a period, and may have deteriorated.
- Normally a change in fluid density is not required to circulate out workover kicks as opposed to drilling kicks.

WORKOVER BLOWOUT EXPERIENCE

The workover blowout experience from January 1980 - January 1994 in the U.S. GoM OCS and the North Sea (UK and Norway) is presented in this chapter and is based on the SINTEF Offshore Blowout Database (see *SINTEF Offshore Blowout Database* on page 33). A total of 19 workover blowouts has been recorded.

Figure 8.1 shows the annual workover blowout frequency related to the number of wells in production each year and the associated regression line.

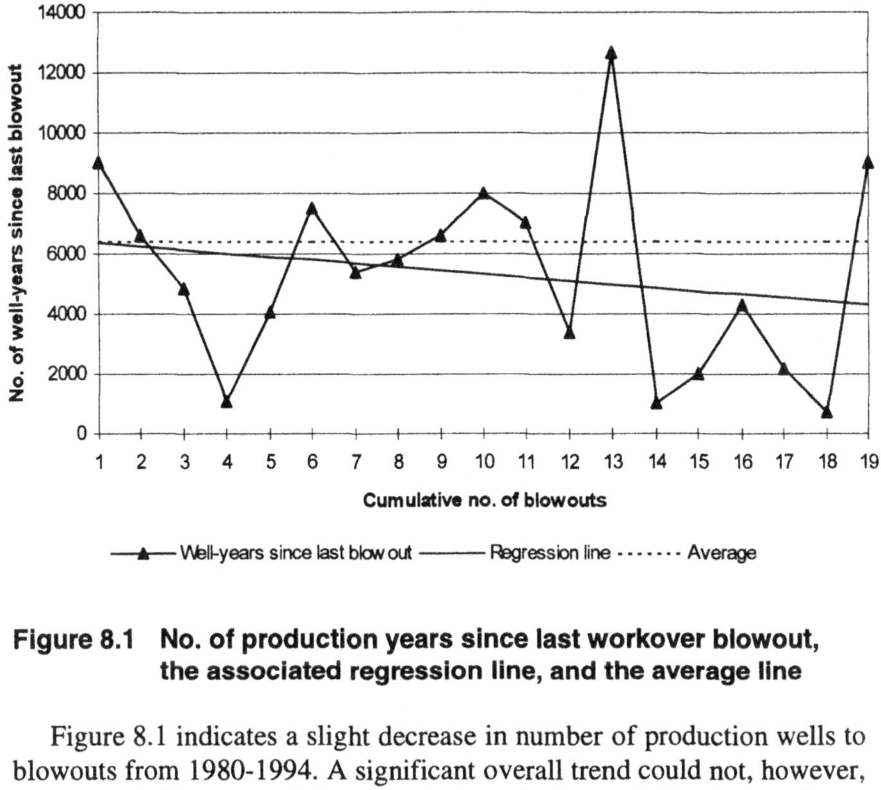

Figure 8.1 No. of production years since last workover blowout, the associated regression line, and the average line

Figure 8.1 indicates a slight decrease in number of production wells to blowouts from 1980-1994. A significant overall trend could not, however, be concluded by any of the statistical methods used. For the workover blowout occurrences, approximately 21,000 production well-years were recorded after the last blowout occurred. Regression analysis does not handle censored data; therefore, the average line lies above the regression line.

Table 8.1 shows the operations and activities in progress when the workover blowouts occurred.

Table 8.1
Operations and Activities in Progress when the Workover Blowouts Occurred

Operation ⇒ Activity ⇓	Pulling well equipment	Installing equipment	Perforating	Well testing incl. preparations	Circulating	Snubbing	Abandon well	Unknown	Total
Tripping out	2	-	-	-	-	-	-	-	2
Pull tubing	5	-	-	-	-	-	1	-	6
Out of hole	-	-	-	1	-	-	-	-	1
Circulating	2	-	-	-	1	-	-	-	3
Cement squeeze	-	-	1	-	-	-	-	-	1
Set well plug	-	1	-	-	-	-	-	-	1
Perforating	-	1	-	-	-	-	-	-	1
Snubbing in	-	-	-	-	-	1	-	-	1
Changing equipment	-	1	-	-	-	-	-	-	1
Unknown	-	-	-	-	-	1	-	1	2
Total	**9**	**3**	**1**	**1**	**1**	**2**	**1**	**1**	**19**

As can be seen from Table 8.1, pulling tubulars from the well was the only repetitive activity when the workover blowouts occurred. Otherwise, two blowouts started during snubbing operations and two blowouts occurred during well circulation.

WORKOVER BLOWOUT CAUSES

The causes for losing the primary and secondary barriers during workovers are discussed in association with Table 8.2. This table lists the causes for losing well control.

Table 8.2
Causes of Losing the Well Control During Workovers

Loss of primary barrier ⇒ Loss of secondary barrier ⇓	Too low hydrostatic head								
	Swab-bing	Too low mud weight	Trap-ped gas	Cement pre-flush weight too low	Un-known why	Snubbing equip-ment failure	Packer plug failure	Poor cement	Total
String safety valve not available	-	-	-	-	1	-	-	-	1
Failed to stab kelly valve	3	-	-	1	-	-	-	-	4
String safety valve failed	1	-	-	-	-	-	1	-	2
Failed to close BOP	2	1	-	-	-	-	-	-	3
Insufficient frictional back pressure	1	1	3	-	1	1	-	-	7
Annulus valve failed	-	1	-	-	-	-	-	-	1
Casing head failed	-	-	-	-	-	-	-	1	1
Not relevant	-	-	-	-	-	1	-	-	1
Total	**7 (6)***	**3**	**3**	**1**	**2**	**2**	**1**	**1**	**20 (19)***

* Figures in parentheses denote number of blowouts. For some blowouts, two secondary barrier failures are reported.

Loss of the primary barrier

Swabbing

For six of the workover blowouts, *swabbing* was the cause of losing the primary barrier. The first swabbing incident occurred when pulling the string out of the hole. There was little information about the second *swabbing* blowout. It seems likely that the kick occurred during pulling out of the well. The third *swabbing* incident occurred during pulling of the packer and tubing when the tubing parted. The fourth *swabbing* incident occurred when pulling the tubing with hydraulic jacks. In the fifth *swabbing* incident, the crew had just pulled the tubing out of a downhole hanger when the well started to flow. In the last *swabbing,* workers had just performed a drill stem test and had removed the drillstring and test tools. After 10-15 minutes the well started bubbling on the bell nipple.

Too low mud weight

Three blowouts were listed with *too low mud weight* as the primary barrier failure cause. The first *too low mud weight* incident occurred after squeezing cement and perforating. The well started to flow gas and cement.

Because of a communication error, the second *too low mud weight* incident occurred when the tubing hold-down pins were removed before killing the dual tubing strings with fluid.

In the third *too low mud weight* incident, workers were circulating 1.5 m^3 (10 bbl) of water down the tubing and out the annulus when the well started to flow.

Trapped gas

For three of the workover blowouts, *trapped gas* was the cause for losing the primary barrier. For the first *trapped gas* incident, the crew had run a tool and had inspected for trapped gas before the incident, but they could not locate any. When trying to release the packer, the tubing broke at 600 m (1,970 ft) and trapped gas blew approximately 215 m (700 ft) of tubing out of the hole.

The second *trapped gas* incident occurred when a blind back off was performed. The trapped pressure beneath the downhole safety valve (SCSSV) forced 133 m (436 ft) of tubing, drillpipe, and drill collar out of the well.

The third *trapped gas* incident occurred because pressure was trapped below the packer. The packer had been released with some problems, and the crew was pulling the tubing hanger assembly when the tubing string started to come out of the hole. A total of 59 m (194 ft) of tubing string came out of the hole before the shear ram preventer was closed.

Cement preflush weight too low

The *cement preflush weight too low* incident occurred because the fluid in the work string was underbalanced during cementing operations. A cement plug had just been set and two stands had been pulled when the incident occurred.

Too low hydrostatic head - unknown why

Little information exists about these two incidents, but the wells should have been hydrostatically controlled when the incidents occurred.

Snubbing equipment failure

Two of the blowouts occurred during snubbing operations. For the first one, a bull plugged outlet on the lower set of rams failed.

A wrong measurement of the snubbing tool joint caused the second incident. The tool joint was believed to be 0.5 m (1.5 ft) shorter than the actual length. This caused the tool joint to be run in the stripper ram. The 1-in. snubbing string buckled and weakened, and the well started to blow.

Packer plug failure

A pump open plug in the packer and surface equipment controlled the well during workover operations. During displacement of mud to low density packer fluid, the well began to flow because the shear pin capacity was accidentally exceeded, which caused the plug to lose integrity.

Poor cement

While reperforating, the casing hanger anchor bolts began leaking. Two 2-in. anchor bolts and gland assemblies were removed so the leaks could be repaired without securing the well. The well began to flow through the screw holes.

Loss of the secondary barrier

Secondary barriers were lost seven times because of string safety valve problems, 12 times because of BOP difficulties, and one time because of a problem with the annulus valve. For the incident reported under *not relevant,* there were no secondary barriers.

String safety valve not available

During workover operations, the well kicked while changing the inside BOP to a tubing safety valve. After a few hours, the blind-shear rams were closed to stop the flow.

Failed to stab kelly valve

The first *failed to stab kelly valve* incident occurred when the crew attempted to stab the valve, but the force of escaping fluid was too great.

For the second *failed to stab kelly valve* incident, the well started to flow slowly through the string. The drilling crew, however, failed to stab the valve because they were not prepared for stabbing in the cement work string. A crossover was needed to fit the stabbing valve to the cement work string, but at the time the crossover was fitted, the flow was too large. Workers also complained that the threads were too fine.

For the third *failed to stab kelly valve* incident, the drilling crew began to run the string in the well because the well started bubbling on the bell nipple. Then the well started flowing through the tubing. Five attempts to install a full opening valve on the tubing failed. After that, the crew *failed to close the BOP* from the remote panel.

The reason for the fourth *failed to stab kelly valve incident* is unclear, but it seems that the well had been flowing through the string for some time before the drilling crew finally managed to stab the kelly valve.

String safety valve failed

For the first of the *string safety valve failed* incidents, the kelly valve and the inline safety valve could not be closed. It was not stated whether the valve could not be turned or if the valve was out of reach.

For the second of the *string safety valve failed* incidents, the valve first leaked in the threads between the valve and the crossover during well circulation. After a while the leakage cured itself. Then the crew decided to shut in the well and attempted to close the kelly valve. This was very hard and they only made half the required turn. Then the valve suddenly started to leak out of the stem. The crew attempted to fully open the valve again, but failed.

Failed to close BOP

For the first *failed to close BOP* incident, the cementing operation apparently interfered with the BOP operation, preventing complete closure of the BOP.

The second *failed to close BOP* incident was also reported with *failed to stab valve* and is described in association with the *failed to stab kelly valve* incidents.

The third *failed to close BOP* incident occurred when pulling the tubing out of the well. While laying down tubing with the remaining string hanging in the slips, the well started to flow water, sand, and gas. The drilling crew tried to pick up the tubing string with the elevators, so they could drop it in the hole, but this attempt failed. Then they evacuated the rig. They mentioned no BOP closing attempts. It seems likely that the BOP did not include a shear ram preventer, or the crew failed to close the shear ram preventer.

Insufficient frictional back pressure

All incidents in which tubulars were blown out of the well are listed under *insufficient frictional back pressure.* This includes incidents where workers failed to close the BOP or incidents where the BOP was closed, but the friction force was not high enough to keep the tubular in the well.

The first reported incident with *insufficient frictional back pressure* as the reason for secondary barrier failure caused approximately 150 m (500 ft) of tubing to be blown out of the well after the tubing had parted at 180 m (600 ft) before the crew closed the BOP from the remote station and shut in the well.

For the second incident with *insufficient frictional back pressure* as the secondary barrier failure cause, the 1-in. snubbing pipe was blown out of the hole, and the gas was ignited by abrasive action. The operator could not reach the BOP control due to heat, so he closed the master valve.

The third *insufficient frictional back pressure* incident occurred in just a few seconds when nine lengths of 2 3/7-in. tubing were blown out of the hole. The shear rams and the master valve were then closed.

In the fourth *insufficient frictional back pressure* incident, trapped gas blew 215 m (700 ft) of tubing overboard before the well was shut in and killed with mud.

The fifth *insufficient frictional back pressure* incident occurred when the tubing hold-down pins were completely backed out prior to killing the tubing strings with fluid. The tubing hanger, tubing strings, subsurface ball valves, and subsurface hanger locator were blown out of the hole from a depth of 90 m (300 ft). After the tubing was blown out, the well was closed with the blind-shear rams and the well was killed with mud.

The sixth *insufficient frictional back pressure* incident occurred when the crew performed a blind back off and trapped pressure beneath the

DHSV, forcing 133 m (436 ft) of tubing drillpipe and drill collars from the well. The total footage was lost overboard.

The last *insufficient frictional back pressure* incident occurred when the driller began to pull the tubing hanger assembly, and the tubing strings began to come out of the hole. The driller closed blind-shear rams to stop the pipe movement. However, 59 m (194 ft) of the dual tieback configuration was forced out of the hole before the tubing strings were cut by the shear rams.

Annulus valve failed

The crew chemically cut the tubing at 120 m (390 ft) to change out the broken SCSSV in the well. Then they opened the annulus valve and circulated 1.5 m^3 (ten bbl) of water into the tubing and out the annulus valve. A flow came out of the crown valve, which the crew subsequently closed and the flow came out of the annulus valve. The valve stem snapped when the crew attempted to tighten the annulus valve.

Casing head failed

The incident listed with *casing head failed* started during a reperforating operation. At first, the casing hanger anchor bolts began leaking. Without first securing the well, two 2-in. anchor bolts and gland assemblies were removed to repair the leaks. The well began to flow through the screw holes. The casing valve was opened to relieve the pressure and to reinstall the bolts, but the flow was too high.

Not relevant

For the incident listed with *not relevant* as the secondary barrier failure, only one barrier was present at the time of the incident. The incident occurred during snubbing operations. When the BOP started to leak through a bull plugged outlet, there were no barriers left.

WORKOVER BLOWOUT CHARACTERISTICS

This section describes the experienced blowout characteristics that are relevant for risk evaluation.

Blowout flow paths and release points

Table 8.3 lists the final flow path vs. release point for workover blowouts.

Table 8.3
Release Point vs. Final Flow Path for Workover Blowouts

Final flow path ⇒ Release point ⇓	Through drillstring	Through tubing	Through annulus	Through outer annulus	Total
BOP valve outlet (snubbing BOP)	-	-	1	-	1
Drill floor - through rotary	-	-	8	-	8
Drill floor - top of drillstring	3	-	-	-	3
Drill floor - top of tubing	-	3	-	-	3
Drill floor - tubing valve	-	1	-	-	1
From wellhead	-	-	-	1	1
From Christmas tree	-	-	1	-	1
Unknown	-	-	1	-	1
Total	**3**	**4**	**11**	**1**	**19**

As seen from Table 8.3, workover blowouts do not have their final flow path outside the casing. The normal flow paths are either through the drillstring/tubing or through the well annulus. Further, blowouts through the drillstring/tubing are mostly released from the top of the drillstring/tubing, which hang in the rotary table slips. Blowouts through the well annulus mostly come through the rotary table.

Blowout flow medium

Table 8.4 shows an overview of the blowout flow mediums for the workover blowouts.

Blowouts, which are recorded with both gas and oil as flow mediums, are listed as oil blowouts, and blowouts with both gas and condensate are referred to as condensate blowouts.

Table 8.4
Blowout Flow Mediums for North Sea and U.S. GoM OCS Workover Blowouts

PHASE	Blowout flow medium			Total
	Gas	Condensate	Oil	
Workover	13	1	5	19

The relative number of oil blowouts is far higher for workover blowouts than for "deep" drilling blowouts (Table 5.5).

Blowout flowrates

Experienced flowrates are usually not reported. A flowrate figure is listed for only one workover blowout.

Flowrates are important figures in risk and environmental analyses. In order to establish a realistic distribution of flowrates to be expected for specific fields, field specific productivity data should be compared with blowout experience with respect to remaining restrictions in the wells during the blowout situations. It is important to consider the fact that well productivity declines with time. Further, for most workover blowouts there are significant flow restrictions that will reduce the flow.

Blowout pollution

None of the workover blowouts caused severe pollution. Six of the nineteen blowouts produced *small* pollution. The most severe pollution was from an oil blowout, which emitted 10 m^3 (63 bbl) into the ocean, and from a condensate blowout, which caused a large sheen.

Out of the 41 workover blowouts worldwide after January 1970, only two blowouts were reported with large spills. One of these blowouts was the well-known 1977 Bravo blowout. Approximately 20,000 m^3 (125,000 bbl) of oil spilled into the North Sea (Norwegian Continental Shelf) during the eight days of the blowout. The oil dissolved and caused no onshore damage. The second one was in 1992 in the U.S. Timbalier Bay's shallow waters offshore Louisiana. Approximately 500 m^3 (3,100 bbl) of oil spilled. The oil drifted ashore and caused damage to wildlife.

Severe pollution incidents may result from a workover blowout, but U.S. GoM OCS and North Sea experience shows that blowouts seldom cause severe pollution incidents in these areas.

Ignition of blowouts

Table 8.5 lists the experience related to workover blowout ignition.

Table 8.5
Experienced Ignition for Workover Blowouts

PHASE	No ignition	Immediate ignition (< 5 min)	Delayed ignition 5 min - 1 hr	1 hr - 6 hr	6 hr - 24 hr	> 24 hr	Total
Workover	14 73.7%	2 10.5%	-	-	-	3 15.8%	19

As seen from Table 8.5, 74% of the workover blowouts did not ignite. Two (10.5%) blowouts ignited immediately. One ignited after 24 hours, one after 36 hours, and another ignited 72 hours after the blowout started.

The overall trends regarding ignition of blowouts are decreasing (see *Ignition Sources and Trends* on page 51).

Blowout duration

Table 8.6 presents the experienced workover blowout durations. Four blowouts were controlled by pumping mud, another four were contained by bridging, six blowouts were restrained by the BOP, and the last five by other topside mechanical devices.

Table 8.6
Workover Blowout Duration

PHASE	Blowout duration ≤ 10 min	10 min - 40 min	40 min - 2 hr	2 hr - 12 hr	12 hr - 5 days	> 5 days	Unknown	Total
Workover	3 15.8%	2 10.5%	-	1 5.3%	7 36.8%	3 15.8%	3 15.8%	19

Casualties

One workover blowout caused two fatalities. This blowout ignited immediately. Otherwise, no casualties occurred.

Material losses

Due to lack of information, a detailed overview of the severity of the various blowouts is difficult to establish. Table 8.7 gives a coarse overview of the workover blowout severity.

Table 8.7
Workover Blowout Severity

Consequence class	No. of blowouts
Small	14
Severe	2
Total loss	1
Unknown	2
Total	**19**

All blowouts cause economic losses, even if there is no topside material damage. In the best case scenario, only time is lost before the operation is back at the same stage as it was when the incident occurred.

All the incidents that did not ignite are listed with small as the consequence. One ignited blowout totally destroyed the jack-up rig. Two of the blowouts caused severe rig damage, one of which cost (US) $7.6 million in repairs for the submersible rig. For the other incident listed with severe as consequence, both the derrick and the living quarters on the jacket collapsed after a few hours. The fire lasted several days. The repaired platform was in production nine months later.

CHAPTER NINE

Production Blowouts

Production blowouts occur from production or injection wells, which may be in service (producing/injecting) or closed in by mechanical well barriers.

For a blowout to occur in a production well, at least one primary and one secondary barrier have to fail. During production both the *primary and secondary barriers are mechanical barriers* (see *Barriers in Well Operations* on page 17). This is therefore different from drilling, workover, and completion blowouts, where the primary barrier is usually the hydrostatic pressure from the mud column.

The SINTEF Offshore Blowout Database includes 12 production blowouts from January 1980 till January 1994 in U.S. GoM OCS and the North Sea (Norwegian, UK).

Out of these 12 blowouts, external forces "caused" six. External forces did not cause blowouts for the other operational phases (drilling, completion, workover, and wireline) in the U.S. GoM OCS and the North Sea for the stated period. The remaining six production blowouts originated from "normal" causes.

BLOWOUTS WITH EXTERNAL CAUSES

An external force normally only damages the topside barrier. For a blowout to occur, the downhole barrier also has to fail. Consequently, an external force will not be the single blowout cause. Typically the external force damages the wellhead/Christmas tree barriers of an active well, and the downhole barrier fails to activate, or leaks, causing an uncontrolled flow.

These blowouts have not been studied in detail. In the more peaceful parts of the world, they normally occur in shallow waters and are caused by a ship collision or a storm. In other parts of the world, such blowouts have occurred because of military attacks. Several such blowouts occurred in the Persian Gulf during the Iranian/Iraqi war.

Table 9.1 shows an overview of blowouts caused by external forces.

Table 9.1
Blowouts Caused by External Forces

Water depth (meters)	Operation	Activity	External cause	Primary barrier	Secondary barrier	Flow medium
6	Closed in gas well	Well closed in	Ship collision	SCSSV/storm choke failure (not enough surge to close valve)	Christmas tree failed (leakage between tubing head flanges and master valve)	Gas (deep)
143	Producing oil	Gas lifting	Fire/-explosion	SCSSV/storm choke failure and a tubing to annulus leakage, equipment or nipple failure (5 to 6 wells failed)	Christmas tree failed (all trees failed due to topside fire)	Oil, Gas (deep)
12	Producing gas	Regular production	Ship collision	SCSSV/storm choke failure (or not installed)	Christmas tree failed (due to collision)	Gas (deep)
10	Producing oil	Regular production	Storm	SCSSV/storm choke failure (assumed, may also have been a tubing annulus leakage)	Christmas tree failed (damaged by storm)	Oil
13	Producing oil	Regular production	Storm	SCSSV/storm choke failure (assumed, may also have been a tubing annulus leakage)	Christmas tree failed (damaged by storm)	Oil
10	Producing oil	Regular production	Storm	SCSSV/storm choke failure (assumed, may also have been a tubing annulus leakage)	Christmas tree failed (damaged by storm)	Oil

Only one of these blowouts occurred in fairly deep waters. This blowout was a result of the Piper Alpha gas leak and fire. Several wells were leaking oil due to downhole failures after the topside structure was totally ruined by the fire.

The remaining five incidents were in shallow waters in the U.S. GoM OCS. It is likely that storms cause more of this type of blowout, but are not reported. The above three blowouts were all detected by helicopters that flew over the areas after a severe storm. The pilots detected oil slicks, which led them to spot the blowouts. Pollution from these incidents was reported as medium. MMS does not keep these incidents in their blowout files.

In addition to the above six blowouts, the database lists 12 worldwide production blowouts, caused by external forces, which occurred after January 1970 because of external causes. Three were caused by storms, two by fire/explosion, six by military attacks, and one by a ship collision.

Three of these 12 blowouts caused severe pollution. Military attacks in the Persian Gulf in 1983 caused two blowouts with severe pollution. In 1970 a fire on an installation in the U.S. GoM OCS caused a blowout with severe pollution.

PRODUCTION BLOWOUT CAUSES

Blowouts with external causes are not included in the subsequent text. Table 9.2 shows the operations and activities in progress when the production blowouts occurred.

Table 9.2
Operations and Activities when Production Blowouts Occurred

Operation ⇒ Activity ⇓	Producing oil	Producing gas	Closed in gas well	Unknown	Total
Regular production	1	3	-	1	5
Failure diagnosing	-	-	1	-	1
Total	**1**	**3**	**1**	**1**	**6**

Since there are only a few blowouts listed, specific trends in operations and activities carried out when the incident occurred cannot be identified. One gas well was closed in due to a tubing to annulus leak. The well had been closed in for some time. Therefore, it was *not* categorized as *regular production* when the incident occurred.

Table 9.3 lists the causes of losing the primary and secondary barriers.

Table 9.3
Loss of Primary and Secondary Barriers for Production Blowouts

Loss of primary barrier ⇒ Loss of secondary barriers ⇓	Poor cement	Tubing to annulus leakage	SCSSV/storm choke failure	Total
Christmas tree failed	-	1	2	3
Annulus valve failed	-	1	-	1
Wellhead seal failed	1	-	-	1
Casing leakage	-	1	-	1
Total	**1**	**3**	**2**	**6**

The well listed with *poor cement* and *wellhead seal failed* started to flow gas, mud, and water between the 13 5/8-in. casing and 20-in. conductor. The cause of the flow was believed to be gas in shallow sand.

For three incidents, tubing to annulus leakage was reported as the primary barrier failure. The secondary barrier failure for one of these blowouts is listed as a Christmas tree failure. It was actually the 3/4-in. test port for the tubing hanger that leaked. The annulus pressure was bled off, and a nipple and valve were installed in the test port. The second incident is from a subsea well. It seems, from a scarce source description, that the annulus valve failed and allowed the well fluids to enter the surroundings. The well was killed with mud and repaired. In the third incident reported with tubing to annulus leakage, a downhole casing failure caused an underground blowout before it resulted in a surface blowout with a crater. The platform tilted after the incident.

The two remaining incidents were reported with SCSSV/storm choke failure. In one incident, workers were not able to close the two master valves. The gas flowed through a needle valve. The well was killed with mud. In the other incident, it took workers 36 hours to close the bottom master valve.

PRODUCTION BLOWOUT CHARACTERISTICS

Blowout flow paths and release points

Table 9.4 lists the final flow path vs. the release points for blowouts during production activities.

Table 9.4
Release Point vs. Final Flow Path for Production Blowouts

Final flow path ⇒ Release point ⇓	Through tubing	Through annulus	Through outer annulus	Outside casing	Total
From wellhead	1	1	1	-	3
From Christmas tree	1	-	-	-	1
Subsea crater	-	-	-	1	1
Subsea Christmas tree	-	1	-	-	1
Total	**2**	**2**	**1**	**1**	**6**

Five of the six blowouts blew out in the wellhead/Christmas tree area. The last blowout came outside the casing at the sea bottom and caused a subsea crater.

Blowout flow medium

Five of the six blowouts were listed with *gas* as the flow medium, one of these five was shallow gas. The last blowout was listed with both *oil* and *gas.*

Blowout pollution

One of these six blowouts was listed with *small* pollution, the other with *no* pollution.

In addition to the above six blowouts, the database lists 11 worldwide production blowouts that occurred after January 1970 (not incl. blowouts with external causes). Of these blowouts, two caused large pollution, one off Trinidad in 1973 and one in the Caspian Sea in 1989.

Ignition of blowouts

None of the six blowouts ignited.

Blowout duration

Table 9.5 presents the production blowout durations.

Table 9.5
Production Blowout Duration

PHASE	Blowout duration							
	< 10 min	10 min - 40 min	40 min – 2 hr	2 hr – 12 hr	12 hr – 5 days	> 5 days	Un-known	Total
Production	-	-	-	-	5 83.3%	-	1 16.7%	6

Production blowouts last relatively long. Two blowouts lasted one day, one lasted one and a half days, one persisted for two days, and another continued for three days.

Casualties

None of the production blowouts involved casualties.

Material losses

The only blowout listed with severe topside damage caused a subsea crater and the platform to tilt. The operator had a (US) $220 million insurance claim after the incident. Otherwise, no severe topside damage was reported. However, such incidents cause production delays and thus may be very costly.

CHAPTER TEN

Wireline Blowouts

Wireline blowouts occur during wireline operations in production or injection wells. Wireline operations are also frequently performed during well workovers, well drilling or well completions. Blowouts that occur during these operations are not regarded as wireline blowouts.

During wireline operations, a stuffing box/lubricator and/or a wireline BOP located on top of the Christmas tree is normally the primary barrier. If the well cannot be controlled by those means, the wireline is dropped or cut before the Christmas tree is closed to control the well.

The SINTEF Offshore Blowout Database includes three wireline blowouts from January 1980 till January 1994 in U.S. GoM OCS and the North Sea (Norwegian, UK waters). There is reason to believe that several more minor blowouts or gas leakages have occurred during wireline operations, but were never reported in public files or articles.

All the wireline blowouts caused small or no damage. However, the quality of the source material is of rather low quality.

WIRELINE BLOWOUT CAUSES

Table 10.1 shows the operations and activities in progress when the wireline blowouts occurred.

Table 10.1
Operations and Activities when Wireline Blowouts Occurred

Operation ⇒ Activity ⇓	Running wireline operations
Pull wireline	1
Run wireline	2
Total	**3**

Since there are few occurrences, specific trends in operations and activities carried out when the incident occurred cannot be identified.

Table 10.2 lists the causes for losing the primary and secondary barriers.

Table 10.2
Loss of Primary and Secondary Barriers for Wireline Blowouts

Loss of primary barrier ⇒ Loss of secondary barriers ⇓	Wireline lubricator failure	SCSSV failure	Total
Christmas tree failed	2	-	2
Wireline lubricator not installed	-	1	1
Total	**2**	**1**	**3**

Two incidents were listed with *wireline lubricator failure* as the primary barrier failure and *Christmas tree failed* as the secondary barrier failure. One blowout occurred when pulling a plug out of the well with a wireline. Then the plug became loose and blew up in the well. The plug then became stuck in the Christmas tree, making closing of the master valve impossible. The wireline lubricator was not mentioned in the blowout description, but using such a device is mandatory [15]. It is therefore assumed that the lubricator failed to seal off the well. The well was killed by pumping mud through the wing valve. In the second incident it was stated that a high pressure lubricator connection broke. The Christmas tree was not mentioned in the description of the incident. Nevertheless, it seems likely that the crew was not able to close the Christmas tree master valve, either because the valve could not cut the wire, or the valve itself failed to close.

The incident in which an SCSSV failure was reported as the primary barrier failure occurred during wireline operations when the master valve was used as a holding clamp for a gas-lift valve that should replace a gas-lift dummy valve. The master valve was opened by mistake and the gas-lift valve fell down on the SCSSV that broke. Fragments were blown out of the well, and two workers were injured before the master valve was closed, which took approximately one minute.

WIRELINE BLOWOUT CHARACTERISTICS

Blowout flow paths and release points

Wireline blowouts typically flow through the tubing. The release point is above the Christmas tree. This was stated for two of the blowouts. For the last one the flow path and release point were not mentioned, but it is likely it was the same as the above.

Blowout flow medium

Two of the blowouts were listed with *gas* as the flow medium. The last one was listed with *oil*.

Blowout pollution

None of the blowouts caused pollution. After January 1970 there were, besides the above three wireline blowouts, three other wireline blowouts worldwide listed in the database. Of these blowouts, one caused extensive pollution. This incident occurred in U.S. GoM OCS in 1970. The blowout caused ten additional wells to blow out. In total, approximately 10,000 m^3 (60,000 bbl) spilled into the sea.

Ignition of blowouts

None of the three wireline blowouts that occurred from January 1980 till January 1994 in the U.S. GoM OCS and the North Sea ignited.

Blowout duration

Table 10.3 presents the wireline blowout durations.

Table 10.3
Wireline Blowout Duration

PHASE	Blowout duration							Total
	< 10 min	10 min – 40 min	40 min – 2 hr	2 hr – 12 hr	12 hr – 5 days	> 5 days	Un-known	
Wireline	-	1 33%	-	1 33%	1 33%	-	-	3

Casualties

None of the wireline blowouts involved casualties.

Material losses

None of the blowouts caused severe topside damage.

CHAPTER ELEVEN

US GoM OCS vs. North Sea Blowout Frequencies

INTRODUCTION

Blowout frequencies are important input data to a wide range of quantified risk analyses of offshore systems. Frequency estimates are often based on data from the U.S. GoM OCS and the North Sea. The blowout and exposure experience is more extensive in the U.S. GoM OCS than in the North Sea. Some analysts prefer to regard the data from these two areas as one data set. The frequency estimate will then be far closer to the U.S. GoM OCS experienced frequencies than to the North Sea experienced frequencies. Other analysts prefer to use the average between the U.S. GoM OCS frequencies and the North Sea frequencies. On average, most U.S. GoM blowout frequencies are higher than the North Sea frequencies.

None of these approaches can be labeled as wrong; however, using different approaches for estimating the blowout frequencies makes it difficult to compare different analyses. Further, this may lead some installations to implement risk-reducing means to meet risk acceptance criteria while other similar installations do not.

This chapter looks more closely at the differences between U.S. GoM and North Sea blowout frequencies. Further, basic blowout frequencies to be used as input for North Sea and U.S. GoM OCS risk analyses are proposed.

BLOWOUT FREQUENCY CALCULATIONS

For the data sets in which no trends are observed, blowout frequencies may be calculated. When calculating blowout frequencies, it is assumed that the number of blowouts during a specific time period may be modeled by a homogeneous Poisson process, with blowout frequency λ [38]. The blowout frequency may be estimated by:

$$\hat{\lambda} = \frac{\text{Number of blowouts}}{\text{Accumulated operating time}} = \frac{n}{s}$$

The number of wells drilled is used as the *accumulated operating time* for drilling blowouts. When estimating completion blowout frequencies, the total number of wells completed is used as the *accumulated operating time*. For production blowouts, the total well producing time is used as the *accumulated operating time*. Well workover blowout frequencies are either computed based on the total well producing time or are roughly estimated by the total number of well workovers. The same applies to wireline blowouts: the total well producing time or a coarse estimate for number of wireline operations is used as the *accumulated operating time.*

The uncertainty in the estimate, λ, may be measured by a 90% confidence interval:

- If the number of blowouts, *n*, > 0, a 90% confidence interval is calculated by:

  Lower limit: $\lambda_L = \frac{1}{2s} \chi_{0.95,\,2n}$

  Upper limit: $\lambda_H = \frac{1}{2s} \chi_{0.05,\,2(n+1)}$

- If the number of blowouts, *n*, = 0, a 90% (single sided) confidence interval is calculated by:

  Lower limit: $\lambda_L = 0$

  Upper limit: $\lambda_H = \frac{1}{2s} \chi_{0.10,\,2}$

where $\chi_{\varepsilon, z}$ denotes the upper 100ε % percentile of the Chi-square distribution with z degrees of freedom [38].

The meaning of the 90% confidence intervals is that the frequency is a member of the interval with a probability of 90%, i.e., the probability that the frequency is lying outside the interval is 10%.

DRILLING BLOWOUTS

In Figure 11.1, the blowout frequencies for the U.S. GoM OCS and the North Sea are illustrated with 90% confidence intervals for various drilling blowout types.

As seen from Figure 11.1, the average drilling blowout frequencies for the North Sea are different from the U.S. GoM OCS frequencies. These differences have been tested for statistical significance according to a general test procedure described in Høyland [37].

The results of the tests show that there is a significant difference in development drilling frequency for the U.S. GoM OCS and the North Sea. Further, there is no statistically significant difference in the blowout frequency for exploration drilling.

The difference in the development drilling blowout frequencies between U.S. GoM OCS and the North Sea should be reflected when analyzing risk associated with North Sea installations. Thus, the total frequency should not be applied because that implies that too high blowout frequencies are used as input data.

Since there are relatively few occurrences in the North Sea for some drilling blowout types, there will be extensive uncertainties in the frequency estimate. Relying on the North Sea frequencies alone is therefore not recommended.

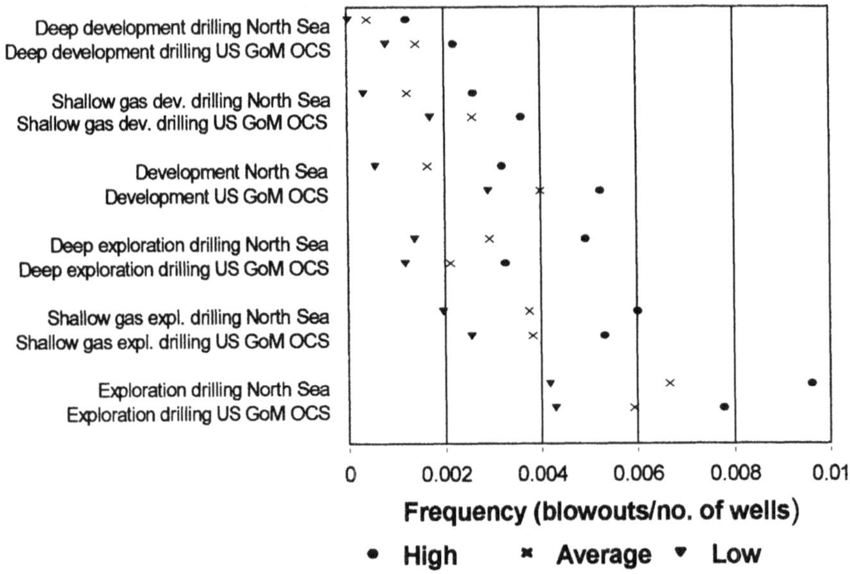

Figure 11.1 Comparison of the drilling blowout frequencies between U.S. GoM OCS and the North Sea, alongside 90% confidence intervals

To consider the statistical uncertainties, but also to give credit for the lower average North Sea frequency, *it is recommended that the average between the U.S. GoM OCS frequencies and the North Sea frequencies be used as input for general risk analysis regarding North Sea drilling blowouts. It is recommended that the U.S. GoM OCS frequencies be used as input for general risk analysis regarding the U.S. GoM OCS drilling blowouts* (See Table 11.1).

WORKOVER BLOWOUTS

In Figure 11.2, the blowout frequencies for the U.S. GoM OCS and the North Sea are illustrated with 90% confidence intervals for workover blowouts.

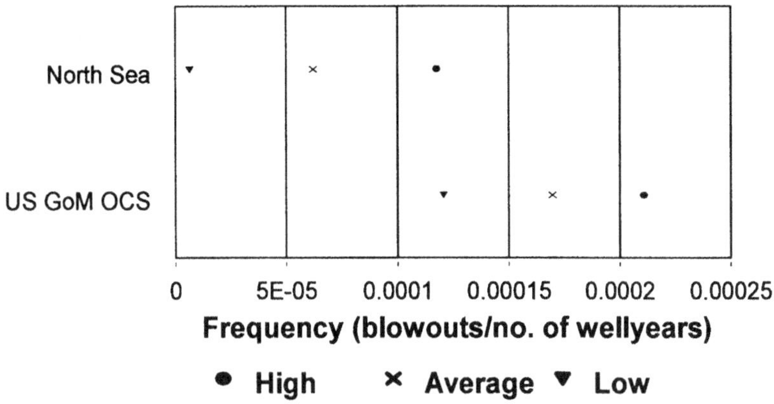

Figure 11.2 Comparison of the blowout frequencies between U.S. GoM OCS and the North Sea for workover blowouts, alongside 90% confidence intervals

There is a large difference in the estimated average blowout frequency between U.S. GoM and the North Sea. Further, the confidence intervals in Figure 11.2 do not overlap.

When testing the two data sets, it was found that the difference was close to being significant. It is, however, important to note that there have probably been more workovers in the U.S. GoM OCS per well-year than in the North Sea. This is because the workover need is higher due to the, on average, older wells in the U.S. GoM OCS. It should further be noted that the North Sea frequency relies on one blowout only. The average U.S. GoM OCS workover blowout frequency is approximately three times higher than the North Sea frequency.

As the average age of the wells in the North Sea increases, the workover need is also likely to increase, which again will heighten the blowout probability per well-year.

When using workover blowout frequencies as input for risk analyses, the frequencies are usually expressed as the number of blowouts per well workover. The expected number of workovers is estimated based on location-specific conditions. An exact count of experienced well workovers does not exist except for certain areas and certain years. Therefore, when using workover blowout experience as input for risk analyses, an estimate for the number of workovers carried out per well-year is used to transfer the data to number of blowouts per workover.

From the SINTEF study, "Reliability of Surface Controlled Subsurface Safety Valves, Phase III", 498 workovers were observed in a total of 7,790 well-years [48]. The data were mainly collected in the period 1985-1989 for North Sea wells. This gives on average:

15.6 well-years per workover

This number is probably too high, because there were many "young" wells included in the data collection.

The *NPD Yearbooks* from 1980 to 1983 list the number of workovers carried out and the number of production wells for North Sea wells. A total of 88 workovers and 731 production well-years were listed. This gives on average:

8.3 well-years per workover

It is normally assumed that the total number of well workovers performed may be expressed by the total number of well-years divided by a number between six and twelve. Since the above workover counts are based on North Sea wells, and U.S. GoM OCS wells on average require more frequent workovers than North Sea wells, it seems reasonable to use a figure closer to six than to twelve. Due to the lack of information regarding the total number of workovers carried out from 1980 until 1994 in the U.S. GoM OCS and the North Sea, it cannot be stated with any confidence that, for instance, one workover per seven well-years is a better estimate than one workover per ten well-years.

It is important that the frequency levels are approximately correct. Further, it is important that various analysts use the same overall frequency as input for various risk analyses. It is therefore proposed that the following is used as a basis for risk analyses - *The average between the U.S. GoM OCS frequency and the North Sea frequency (no. of workover blowouts/number of well-years) is used as input for general risk analysis regarding workover blowouts for North Sea installations. The*

U.S. GoM OCS frequency is used as input for general risk analysis regarding workover blowouts for U.S. GoM OCS installations. Further, in the reference data material it is assumed that, on average, one workover is performed every eight well-years (See Table 11.1).

PRODUCTION BLOWOUTS

Relatively few production blowouts (disregarding blowouts with external causes) are included in the database for the U.S. GoM OCS and the North Sea. Of the six production blowouts, one occurred in the North Sea.

In Figure 11.3 the blowout frequencies for the U.S. GoM OCS and the North Sea are illustrated with 90% confidence intervals for production blowouts.

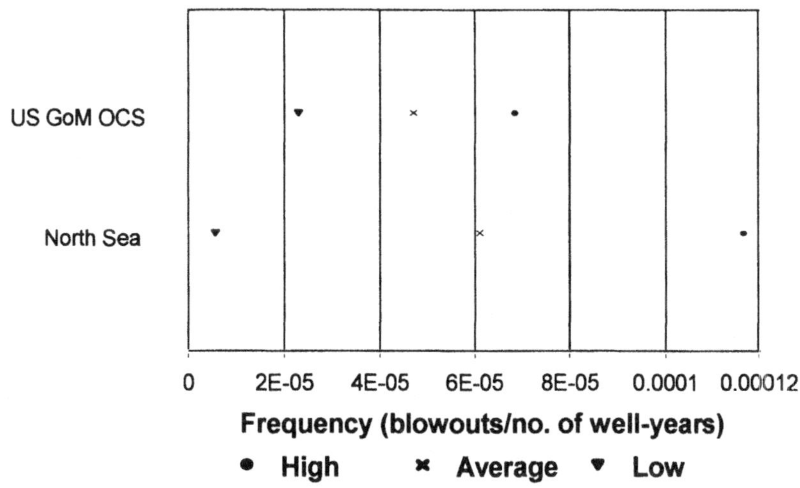

Figure 11.3 Comparison of the blowout frequencies between U.S. GoM OCS and the North Sea for production blowouts, alongside 90% confidence intervals

For production blowouts, the North Sea average frequency is higher than the U.S. GoM OCS average frequency. However, because only a few blowouts occurred in the North Sea, the confidence interval is so wide that the U.S. GoM OCS confidence interval is covered by the North Sea interval.

When testing the production blowout data sets for a statistically significant difference, no significant difference was revealed.

The production blowouts exposure data are defined as number of well-years.

It is proposed that the following is used as a basis for risk analyses - *The average between the U.S. GoM OCS frequency and the North Sea frequency (no. of production blowouts/number of well-years) is used as input for general risk analysis regarding this blowout type for North Sea installations. The U.S. GoM OCS frequency is used as input for general risk analysis regarding this blowout type for U.S. GoM OCS installations.*

WIRELINE BLOWOUTS

Few wireline blowouts are included in the database for the U.S. GoM OCS and the North Sea. All three wireline blowouts occurred in the U.S. GoM OCS.

Regarding the wireline blowouts, very little statistical material related to the number of wireline runs exists.

To establish an estimate for wireline exposure data, experience from the Ekofisk field in 1992 has been used. In 1992, 135 wells were in service (production and injection), and 220 wireline jobs were performed. If, on average, each wireline job includes 2.5 wireline runs, then 550 wireline runs were carried out for the 135 wells. This gives on average:

4.2 wireline runs per well-year or
1.7 wireline jobs per well-year

It is important to note that the Ekofisk field mainly uses wireline retrievable SCSSVs, not tubing retrievable SCSSVs, which most operators today prefer when completing new wells.

It should further be noted that it is likely that many minor blowouts (small gas releases) occurred during wireline jobs, but were never recorded as blowouts.

It is proposed that the following is used as a basis for risk analyses - *The average between the U.S. GoM OCS frequency and the North Sea frequency (no. of production blowouts/number of well-years and no. of wireline blowouts/number of well-years) is used as input for general risk analysis regarding this blowout type for North Sea installations. The U.S. GoM OCS frequency is used as input for general risk analysis regarding this blowout type for U.S. GoM OCS installations*

COMPLETION BLOWOUTS

Trends in completion blowout probability

Of the seven completion blowouts, *one* occurred in 1980, *five* in 1981 and *one* in 1987. This indicates a trend in the blowout probability, and that the present blowout probability level is lower than the average blowout frequency would suggest. Significant overall trends in blowout frequency for *exploration drilling, development drilling*, and *workovers* could not be identified (Chapters 5, 6 and 8).

The trend in completion blowouts has also been investigated. The completion blowouts and the cumulative number of wells completed have been analyzed by three different statistical trend tests. Both the Laplace Test and the MIL-HDBK Test confirmed that there is a significantly decreasing trend [38].

The blowout frequency as a function of time may be modeled by the following expressions:

Power law model: $\alpha\beta t^{(\beta - 1)}$
where $\alpha = .09630$ and $\beta = .47571$

Log linear model: $EXP(\alpha+\beta t)$
where $\alpha = -5.22387$ and $\beta = -.00077$

Where t represents the cumulative no. of wells completed, and the parameters α and β are estimated by the maximum likelihood principle. Figure 11.4 shows the estimated blowout frequency as a function of t, based on the Power law and Log linear model, respectively, for the interval 1- 8,500 well completions.

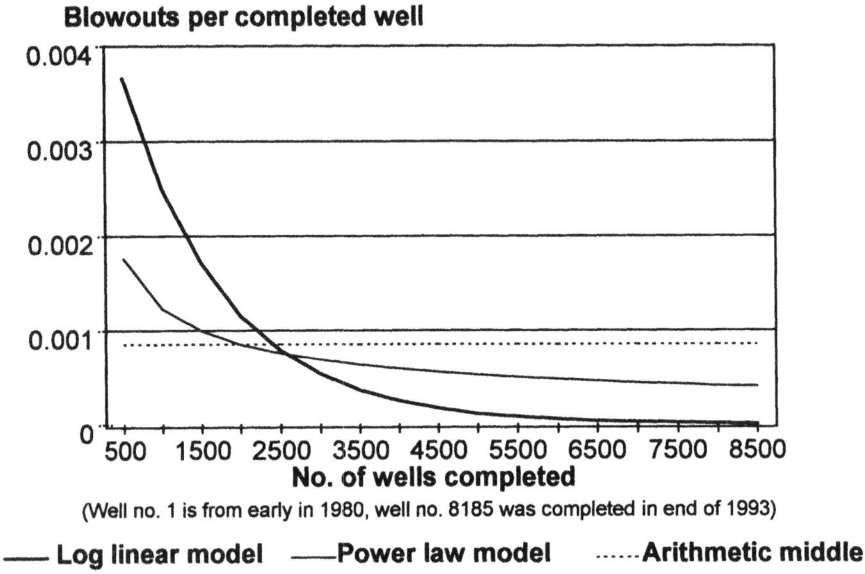

Figure 11.4 Estimated blowout frequency vs. cumulative no. of wells completed

8,185 wells represent the 1993/1994 blowout frequency level. The estimated frequency for 1993/1994, based on the two different estimation methods, will then be:

Log linear model:	0.00001 blowouts per completion,
Power law model:	0.00041 blowouts per completion

The Log linear model seems to give an unrealistically low blowout probability for 1993/1994. The Power law model, which did not fit too well to the actual blowout data, probably gives a conservative result.

As an input for risk analyses, select the final blowout frequency estimate somewhere between these two values. *It is proposed that the average value of 0.00021 blowout per completion is used as the input frequency for risk analyses* (See Table 11.1).

BLOWOUT FREQUENCIES FOR RISK ANALYSES

The previous sections presented various aspects to consider when assessing the basis blowout frequency to be used as input for risk analyses. It is proposed that the U.S. GoM OCS risk evaluations should be based on the U.S. GoM OCS frequency alone. Relatively few blowouts have occurred in the North Sea, so North Sea risk evaluations should be based on both the North Sea and U.S. GoM OCS frequencies. It is therefore proposed that the average of the North Sea and the U.S. GoM OCS frequencies is used as input for the North Sea risk evaluations. Table 11.1 shows the resulting blowout frequencies that may be used as input basis data for risk analyses in the U.S. GoM OCS and the North Sea. These recommended frequencies are based on the author's evaluation and analyses of the blowout data and do not represent an official SINTEF point of view.

Table 11.1
Blowout Frequencies to be Used as Input Basis Data for Risk Analyses of North Sea Installations

Phase		U.S. GoM OCS experienced and recommended frequency	North Sea experienced frequency	North Sea recom-mended frequency	Unit
Explor-ation drilling	Shallow gas	0.00382	0.00375	0.00379	blowouts per well drilled
	Deep	0.00210	0.00291	0.00251	blowouts per well drilled
	Total	0.00593	0.00666	0.00630	blowouts per well drilled
Devel-opment drilling	Shallow gas	0.00257	0.00123	0.00190	blowouts per well drilled
	Deep	0.00142	0.00041	0.00092	blowouts per well drilled
	Total	0.00399	0.00165	0.00282	blowouts per well drilled
Workover		0.00136	0.00050	0.00093	blowouts per well workover *
		0.00017	0.00006	0.00012	blowouts per well-year
Production		0.00005	0.00006	0.00005	blowouts per well-year
Wireline		0.000007	-	0.000003	blowouts per wireline run**
		0.000017	-	0.000008	blowouts per wireline job**
		0.000028	-	0.000014	blowouts per well-year
Completion		0.00021***		0.00021***	blowouts per well completion

* One workover every 8 well-years, see *Workover Blowouts* on page 134

** 4.2 wireline runs per well-year, 1.7 wireline jobs per well-year, see *Wireline Blowouts* on page 138

*** Based on trend analyses, see *Completion Blowouts* on page 139

Glossary

This glossary is a guide for readers who are not familiar with offshore operations. It should be noted that this glossary is related to terms used in the book only, and not a general oilfield glossary of terms. Most of the terms are from Maclachlan, M., *An Introduction to Marine Drilling*, which includes a comprehensive glossary of terms for offshore drilling operations [46].

ANNULAR LOSSES: See lost circulation
ANNULAR PREVENTER: A device in a BOP stack that can seal around irregular-shaped objects, such as drillpipe, that pass through it. It can also close an open hole. It is activated by hydraulic compression of a reinforced rubber or rubber-like packing element.
ANNULUS: The space surrounding any tubular suspended in the hole. During drilling the circulation fluid flows up the annulus between the drillpipe and the wall of the hole, or, when the well is cased, between the drillpipe and the casing. During a casing cement job, a cement slurry is pumped from the bottom of the casing up the annulus between the casing and the wall of the wellbore.
APPRAISAL WELL: A well drilled following the drilling of a discovery well to determine the extent of the oil or gas field. In this book an appraisal well is regarded as an exploration well.
BARITE, BARYTE or BARYTES: Barium sulfate, a mineral used to increase the weight of drilling mud. Its specific gravity is approximately 4.2, i.e., barite is 4.2 times as heavy as fresh water. Quantities of barite in bulk powder form are transported to rigs by supply boats and stored in special tanks until required for mixing with water or oil and other additives to make mud.
BIT: The cutting device used to drill a well. Rotary bits are attached to the bottom of a drillstring which is rotated mechanically, and have nozzles through which the circulating fluid is expelled at high velocity.

There are many types of bits for different geological structures, each type being manufactured in a range of diameters.

BLIND-SHEAR RAMS: Rams in the blowout preventer that can both shear the pipe in the hole and seal off the well.

BLOWOUT: A blowout is an uncontrolled flow of fluids from a wellhead or wellbore. Unless otherwise specified, a flow from a flow line is not considered a blowout as long as the wellhead control valves can be activated. If the wellhead control valves become inoperative, the flow is classified as a blowout.

BLOWOUT PREVENTER: A device to control formation pressures in a well by sealing the annulus around the drillpipe when pipe is suspended in the hole, or alternatively by sealing across the entire hole if no pipe is in it. Another type can shear the drillpipe passing through the preventer. Different types of preventers are assembled in a blowout preventer (BOP) stack.

BOTTOM HOLE ASSEMBLY: The assembly of heavy drilling tools made up in the lower part of the drillstring to put weight on the bit and keep the drillpipe above in tension. In addition to the bit, the *BHA* normally includes drill collars, stabilizers, reamers, heavy weight drillpipe and other assorted tools.

BRIDGE (DOWNHOLE): An obstruction in the hole usually caused by the wall of the hole caving in or by the entry of a large boulder from the wall.

BRIDGE PLUG: A short cylindrical tool that can be set inside casing to provide a pressure-tight seal to isolate a zone. Bridge plugs are set when squeeze cementing, fracturing, plugging and abandoning, or when testing an upper or lower zone. Afterwards they can be retrieved or drilled through.

CASED HOLE: A drilled hole in which steel casing has been set.

CASING: Steel pipe set in the hole as drilling progresses to line the hole wall, preventing caving-in and providing a passage to the surface for drilling fluid and for hydrocarbons if the well is proved productive.

CENTRALIZER: A device with bowed strips of metal running vertically between two collars and fitted around a joint of casing to contract the wall of the hole and keep the casing centralized. This allows more even distribution of cement.

CHOKE AND KILL MANIFOLD: A large assembly of pipes, valves, and chokes installed on the drill floor for controlling downhole formation pressures in emergencies such as kicks and blowouts.
CHOKE LINE: A line connected to the BOP stack used to direct and control the flow of well fluids from the annulus. On a semi-sub or drill ship, the choke line runs up the marine riser to the choke and kill manifold on the drill floor.
CHRISTMAS TREE: A high pressure assembly of valves, pipes, and fittings installed on a wellhead after completion of drilling to control the flow of oil and gas from the casing.
CONDUCTOR CASING: A short string of large-diameter casing that is run by a floater to keep the top of the hole open and prevent sloughing, to serve as a base for wellhead equipment, and to convey the drilling fluid up to the marine riser. Deep well casing programs may call for two conductor casing strings: an outer 30-in. conductor and an inner 20-in. conductor.
CORE: A sample drilled out of the bottom of the well in a solid cylindrical block and retrieved for examination and analysis.
CRATER, TO: To cave in. In a violent blowout the surface around a well can fall into a large hole blown into the seabed by the force of escaping mud, gas, oil, and water. This can cause a bottom-supported drilling unit to collapse and sink.
CROSS-OVER SUB: A tubular tool with box and pin threads of different diameters so that it can be used to provide the connection between two strings of different gauge pipe.
DERRICK: The tall girdered tower erected over the drill floor that supports the hoist and the drillstring and other tubulars that are run into the hole. Land rig derricks are usually portable, in which case they are called masts, whereas derrick refers to the more permanent type of structure. A typical large semisubmersible's derrick might be 49 m (160 ft) high, with a 40-foot square base.
DEVELOPMENT WELL: A well drilled following the drilling of a discovery well to exploit an oil or gas reservoir. It is usually drilled from a fixed platform.
DIRECTIONAL SURVEY: A survey carried out with downhole instruments to measure the angle and azimuth of deviation of the hole

from the vertical. These surveys are usually either magnetic surveys or, where steel casing has been set, gyro surveys.

DIRECTIONAL WELL: A well drilled at an angle from the vertical for one of a number of reasons.

DIVERTER: A T-shaped pipe attached to the top of the marine riser that closes the vertical passage and directs the flow of well fluids away from the rig floor and overside.

DOWNHOLE DRILLING MOTOR: A tubular tool that converts the hydraulic power in a stream of drilling fluid into rotary power, so as to turn a bit. Sometimes called a mud motor or turbo drill.

DRAWWORKS: The large winch situated on the drill floor, which controls the movement of the hoist and around which the fastline part of the drilling line is wound.

DRILL COLLAR: A length of steel pipe much heavier than the drillpipe, several of which are placed at the bottom of the drillstring just above the bit to add weight to the bit and stiffen the bottom hole assembly. Non-magnetic drill collars, sometimes called monels, are used when magnetic survey tools are to be run.

DRILL FLOOR: The area beneath the derrick in the center of which is the rotary table and from which drilling operations are conducted. Sometimes called the rig floor.

DRILLPIPE: Connected lengths of tubing, usually steel, on the end of which the bottom hole assembly and the drill bit are suspended.

DRILL STEM: The assembly of drillpipe that runs from the kelly to the top end of the bottom hole assembly.

DRILL STEM TEST: A test usually run over a period of several days to determine whether commercial quantities of oil or gas are in the formations drilled through. Drilling may continue after the test period to explore deeper zones, or the well may be completed or plugged and abandoned, depending on the findings.

DRILLSTRING: The assembly of drillpipe and other tools that runs from the kelly down to the bit.

DRILLING BREAK: A sudden increase in the rate of penetration of the bit when it enters a zone of softer material in the formation. This may give advance warning of a kick.

DRILLING FLUID: The fluid circulated down the well and back up to the rig for a number of important purposes, including the containment of information pressure, the retrieval of cuttings, bit lubrication and cooling, plastering the wall of the well, providing a well data source. Usually referred to as mud, although air, gas, and foam can also be used as a drilling fluid.

DRILLING RIG: In offshore jargon, any vessel or machinery and equipment used for drilling a well. Strictly, the term should only apply to the drilling plant onboard, while the platform on which the rig stands should be referred to as the barge, platform, unit, or vessel.

ELEVATOR: A latching device, attached to the hook/traveling block assembly by two long links, and placed around the end of a tubular joint when running into or pulling out of the hole. There are special elevators for casing, drillpipe, drill collars, and tubing.

EXPLORATION WELL: A well drilled to find a new reservoir of oil or gas. It may be drilled in a totally new exploration area, in which case it is termed a wildcat well, or it may be drilled to find a new producing formation in an existing field. In this book appraisal wells are also regarded as exploration wells.

GAS CAP: The free gas that lies above an oil reservoir in a formation.

GAS-CUT MUD: Mud containing bubbles of formation gas, giving it a characteristic fluffy texture. The gas is removed in the de-gasser.

GOING INTO THE HOLE: Running drilling equipment into the hole.

GUIDE BASE: A heavy steel frame placed on the seabed to guide tools into the hole and to serve as a foundation for other equipment such as the wellhead and the BOP stack. The temporary guide base is run before the permanent guide base.

GUMBO: A sticky type of clay sometimes encountered during drilling in certain areas that tends to clog equipment.

HANGING OFF: The operation of landing the drillstring in the wellhead with a special tool and unlatching the lower marine riser package from the BOP stack so that the drilling rig can be moved quickly off location. This is an emergency measure.

HOLE OPENER: A large-diameter bit used to make the initial entry into the seabed. On some types, drilling fluid circulation pressure swings the swiveling cutter arms outward, thereby increasing the bit diameter as it starts drilling.

HYDROSTATIC HEAD: The pressure exerted by the weight of a column of liquid at rest, considered in terms of its height.

JACKET: A structure made from tubular pipe fixed to the seabed to support a platform. Production platform jackets are usually towed out to locations and sunk into position.
JACK-UP RIG: A self-elevating mobile offshore drilling platform.
JOINT: A single length of drillpipe or other type of tubular.
KELLY: A long steel pipe, usually with a hexagonal, or sometimes square cross section, suspended from the swivel and connected to the drillpipe. It transmits torque from the rotary table to the drillstring and is able to move vertically, permitting the gradual lowering of the bit. It is hollow, allowing the passage of drilling fluid for circulation purposes, has box and pin threaded ends, and is usually either 12.2, 14.0 or 16.5 m (40, 46 or 54 ft) long.
KELLY BUSHING: A sliding device, through which the kelly fits closely and engages with the master bushing of the rotary table so that rotary torque can be transmitted to the kelly while simultaneously allowing the kelly to move up or down.
KELLY VALVE: A valve installed between the swivel and the kelly to relieve the swivel and rotary hose from fluid pressure when necessary. Also called kelly cock.
KICK: An unexpected flow of formation fluids into the wellbore.
KILL LINE: A high pressure line attached to the BOP stack through which heavy drilling fluid can be pumped into the hole to kill a well. On a semisubmersible or a drill ship the kill line runs down the side of the marine riser.
LOST CIRCULATION: The loss of quantities of drilling fluid into a formation. This may be due to caverns, fissures or permeability. It is evidenced by lack of returns of drilling fluid and stopped by the pumping downhole of lost circulation materials.
LOWER MARINE RISER PACKAGE (LMRP): An assembly comprised of the flex or ball joint, an annular blowout preventer, hydraulic accumulators, sections of riser and the riser slip joint, all of which can be detached from the rest of the BOP stack in an emergency to allow the drilling unit to move off location whilst leaving the well secure.
MARINE RISER: The large-diameter pipe connecting the BOP stack to the drill floor of a semisubmersible or drill ship through which the drillstring passes to the well and through which returns of drilling fluid pass from the well to the rig.

MEASUREMENT WHILE DRILLING (MWD): A technique of logging certain information about downhole conditions through sensors in the bottom hole assembly. Information is then sent to measuring devices and a digital display on the drilling rig by means of pulses transmitted by telemetry through the mud. This requires rotary drilling to be stopped only for a short period and does not interrupt drilling at all if a downhole motor is being used. It is often used in directional drilling to measure angles of inclination and direction.

MILL: A special tool with a rough, sharp, and very hard cutting head used for milling. Mills are made in many shapes and either fit on the end of a drillstring or are incorporated within it like a reamer.

MILLING: Using a mill to grind down metal debris in the hole, remove sections of casing when sidetracking, or reaming out tight spots in the hole.

MOONPOOL: The void space cut in the deck of a semisubmersible, or inside a drill ship, which is open to the water and through which sub-sea equipment is run. Commonly used to describe the area of the cellar deck immediately around the void. A small moonpool is also used for running diving equipment.

MUD: Liquid drilling fluid circulated down the hole and back to the rig.

PACKER: A tubular sealing device that can be lowered into the casing, liner or open hole and made to expand flexible rings at its circumference in order to isolate a section of the hole (e.g., for well testing purposes). Different designs are made for a variety of uses. Packers generally have a hole through their stems for circulating drilling fluid or for running wireline tools, and they may have box and pin connections for the attachment of other tools.

PERMEABILITY: The ability of hydrocarbons to flow through the pores of a rock.

PIN CONNECTOR: A pressure-sealed device used to connect the marine riser to the wellhead when drilling through large diameter conductor casing.

PITS: Large tanks in which drilling fluid is held prior to pumping down the well by the mud pumps. They are usually located near the mud pumps.

PLATFORM: The name sometimes used to describe a mobile drilling unit on which a drilling rig is erected. The term is also commonly used offshore to describe a self-contained fixed production platform.

PLUG: A device inserted into a drilled hole to block the passage of fluids. Plugs may be made of rubber, cement, or other substances, and some can be drilled out or retrieved when no longer required.

RAM: A closing and sealing device in a BOP stack. The rams are activated by hydraulic pressure when a blowout threatens, and can be locked shut.

REAMER: A downhole tool sometimes included in the bottom hole assembly, which looks like a drill collar with short fines on which there are cutters. It is used to smooth and enlarge the wall of the hole, stabilize the bit, straighten the hole where doglegs occur, and it is also used in directional drilling. If the cutters revolve, the tool is called a roller reamer.

REVERSE CIRCULATION: The circulation of the drilling fluid opposite to the normal direction (i.e., down the annulus around the drillpipe and up through the center of the drillpipe). This is sometimes done to alleviate problems in the hole.

RIG: Strictly speaking, the derrick and drilling equipment that are mounted on a platform (semisubmersible, drill ship, etc.). In practice, however, the drilling unit itself is commonly referred to as the rig.

RIG FLOOR: An alternative name often used for the drill floor.

RISER: The steel conduit connecting a rig or platform to the seafloor. A riser is used for production wells and during drilling (see marine riser).

ROTARY BUSHING: A circular steel, cup-shaped lining that fits into the rotary table and into which the kelly bushing is inserted during drilling. When the kelly bushing is removed, slips can be wedged into the space between the rotary bushing and drillpipe running through the rotary. Also called the master bushing.

ROTARY DRILLING: Drilling with a bit that is rotated while a force is applied above it - the normal method of drilling an offshore well. The rotary drive may be applied by the rotary table, topdrive or by a downhole drilling motor.

ROTARY TABLE: The underdeck housing for the mechanism in the center of the drill floor that drives the kelly and turns the drillstring and bit. All downhole tools, casing, etc. are run through its opening.

SATELLITE WELL: A well drilled independently of a platform by a mobile unit, but tied in to the platform for production purposes by a sea-bed pipeline. Most platform wells are directionally drilled from the platform.
SEMISUBMERSIBLE: A type of vessel that, by flooding certain compartments, can be increased to submerge much of its structure, either to give a degree of stability not attainable in conventional monohull vessels, or for some other purpose, such as to float another vessel onto its deck. The hull or hulls may be designed to rest on the sea-bed in certain conditions, but most semisubmersible rigs drill whilst floating.
SHALE: The type of rock most frequently encountered during offshore drilling, composed of small silt and clay particles.
SHALE SHAKER: The vibrating screens across which the drilling fluid returning from the hole is poured to strain off the liquid and deposit the solid particles.
SIDETRACK, TO: To divert the drill bit around an obstruction in the well, such as stuck pipe. This is done using directional drilling techniques and tools such as a whipstock.
SLIP JOINT: A telescopic joint inserted near the top of the marine riser to absorb the vertical heaving motion of the drilling unit when in a seaway.
SLIPS: Tapered steel wedges that are inserted between the rotary bowl and a tubular joint to grip the string temporarily (e.g., while it is disconnected from the hoist when making a connection). The wedges are hinged so that they effectively wrap around the tubular to provide a grip around its circumference. Different types of slips are used for drillpipe, collars, and casing.
SLURRY: A semi-liquid mixture of cement powder and water that is pumped up into the annulus between the casing and the wall of the hole so that it can harden and fix the casing in place.
SPUD IN, TO: To commence drilling a well with a hole opener.
STAB, TO: To insert the pin end of one pipe into the box end of another when making a connection.
STABILIZER: A downhole tool used for stiffening the bottom hole assembly and for keeping the bit centered. It looks like a drill collar but has short fins that contact the wall of the hole. One or more stabilizers

may be used in directional drilling when they are positioned to act as a fulcrum about which the assembly turns.

STACK: The name commonly used offshore for the blowout preventer stack.

STAND: Three or sometimes two joints of pipe screwed together for easier and faster handling on the drill floor.

STAND-PIPE: A tall, rigid pipe in the side of the derrick that carries drilling fluid up from the mud pumps and feeds it into the rotary hose that is suspended between its top and the swivel.

STORM CHOKE: A storm choke is a DHSV (Down Hole Safety Valve) that is flow controlled. See also Surface Controlled Subsurface Safety Valve (SCSSV).

SUBMERSIBLE: A type of drilling platform which is designed to be floated to its location and then sunk so that its bottom rests on the seabed.

SURFACE CONTROLLED SUBSURFACE SAFETY VALVE (SCSSV): A SCSSV is located in the production tubing subsurface. The valve can be used for closing in a well if a topside situation occurs that disables the Christmas tree valves. The valve is controlled from the surface. These valves are frequently referred to as DHSVs (Down Hole Safety Valves). A DHSV does, however, not have to be surface controlled; it can be flow controlled. The flow controlled valves are frequently referred to as storm chokes.

SWABBING: Swabbing is the action of creating a suction in the wellbore which may induce well fluids out of the formation, creating a kick. Swabbing is usually caused by pulling the drillstring too quickly out of the well.

SWIVEL: The device which hangs from the hook below the traveling block that permits free rotation of the kelly, whilst at the same time admits drilling fluid to it from the rotary hose.

TOOL JOINT: A short section of special steel pipe welded around each end of a joint of drillpipe to provide a means of connection and lifting. There are shoulders on the tool joints that the elevators grip when lifting.

TOP DRIVE: A type of drilling swivel that is turned by electric or hydraulic power and replaces the rotary table, master bushing, kelly, and circulating swivel.

TRAVELING BLOCK: The lower, movable block of the hoist, which is suspended by the drilling line from the upper, or crown block.
TRIPPING: The operation of pulling the drillstring out of the well or running the drillstring into the well.
TUBING: Narrow-bore pipe which is run down through the casing to serve as a channel for produced oil and/or gas.
UNDERREAMER: A downhole tool with rock bit cones on the ends of pivoting arms that can expand from the sides of the tool to open up a previously drilled hole. This might be done to provide extra clearance for running casing so as to obtain adequate annular space for cementing, etc.
WAITING ON CEMENT: A period of several hours that must elapse after a cement job to allow the cement to set. No downhole work is done during this time.
WASH OUT, TO: To erode a metal object, such as a drilling tubular joint or a valve, by the action of fluid pressure. Washing out of damaged tool joints may occur through leakage.
WELL COMPLETION: The final phase of operations after total depth has been reached (e.g., when the well is fitted with production equipment).
WELLHEAD: A cylindrical device placed at the top of the hole by a floater in which casing hangers are fitted and sealed and to which well control equipment can be attached during drilling and subsequent production. The BOP stack and, later, the Christmas tree are attached to the wellhead.
WILDCAT: An exploration well drilled in an unproved area, far from any existing producing well.
WIRELINE: A long, narrow wire wound on a storage drum on the drilling rig and used for well logging and well equipment maintenance.
WORKOVER: An operation in which a rig is employed to restore or improve production from a completed well.

References

1 *Accidents Associated With Oil and Gas Operations,* OCS report MMS 92-0058, U.S. Department of Interior, Minerals Management Service, USA, May 1992.

2 *Accidents Associated With Oil and Gas Operations,* OCS report MMS 95-0052, U.S. Department of Interior, Minerals Management Service, USA, September 1995.

3 *Acts, Regulations and Provisions for the Petroleum Activity,* Norwegian Petroleum Directorate, Norway, 1995.

4 Adams, N., and Kuhlman, L., *Kicks and Blowout Control, Second Edition,* Pennwell Publishing Company, Tulsa, OK, USA, 1994.

5 Adams, N. et al., *Shallow Gas Blowout Kill Operations, SPE 21455,* SPE Middle East Oil Show, Bahrain, November 16-19, 1991.

6 *A Guide to the Offshore Installations (Safety Case) Regulations,* 1992, Health and Safety Executive, Sheffield, United Kingdom.

7 API Standard 14J, *Design of Hazards Analysis for Offshore Production Facilities,* American Petroleum Institute, First Edition, Washington D.C., USA, November 1993.

8 *API Standard Definitions For Petroleum Statistics*, Technical Report, No.1, Fourth edition, Washington D.C., USA, 1988.

9 API Standard RP75, *Recommended Practices for Development of a Safety and Environmental Management Program for Outer Continental Shelf (OCS) Operations and Facilities,* American Petroleum Institute, First Edition, Washington D.C., USA, May 1993.

10 Bedford, S. A., and Smith, I., *Coiled tubing operations in the northern North Sea, Magnus Field,* Amsterdam 2nd International Conference and Exhibition on Coiled Tubing, Amsterdam, The Netherlands, June 27-30, 1994.

11 Bird, F. E., and German, G. L., *Practical Loss Control Leadership,* Institute Publishing, Division of International Loss Control Institute, Loganville, GA, USA, 1985.

12 *BOP Stacks and Well Control Equipment Consulting and Training,* 1993 Newsletter, West Hou Inc, Katy, TX, USA, 1993.

13 Borehole list for exploration and development wells, Digitale Datasammenstillinger fra Oljediretoratet, Norway, 1996.

14 *CARA Fault Tree*, SINTEF Industrial Management, Safety and Reliability, Trondheim, Norway, 1996.

15 *Code of Federal Regulations, 30, Parts 200 to 699*, Revised as of July 1995, Office of Documents of General Applicability and Future Effect, Washington D.C., USA, 1995.

16 Corner P.J. et. al., *A Driller's Hazop Method, SPE 15867, EUROPEC*, London, United Kingdom, 1987.

17 *Development of the Oil and Gas resources of the United Kingdom (Brown Book),* Department of Energy, London, United Kingdom, June 1980.

18 *Development of the Oil and Gas Resources of the United Kingdom, (Brown Book)*, 1992 and 1993 editions, Department of Trade and Industry, London, United Kingdom.

19 *EU Common Position, Control of Major Accident Hazards Involving Dangerous Substances (Seveso-Directive),* Brussels, Belgium, 1995.

20 *Fatal Accidents at Work, Fatal accidents in the Nordic Countries over a ten-year period*, The Danish Working Environment Service, Copenhagen, Denmark, April 1993.

21 Fosdick, M., R., *Compilation of Blowout Data from Southeast U.S./Gulf of Mexico Areas Wells*, Master of Science thesis, University of Texas at Austin, August 1980.

22 Grace R. D., *Advanced Blowout & Well Control,* Gulf Publishing Company, Houston, TX, USA, 1994.

23 Grepinet, M. *The Shallow Gas Threat, a Difficult Challenge to cope with*, 6th Annual IBC Tech LTD Offshore Drilling Technology Conference, Aberdeen, Scotland, 25-26 November, 1992.

24 *Guidelines for Chemical Process Quantitative Risk Analysis,* American Institute of Chemical Engineers, AIChe, New York, USA, 1989.

25 *Guidelines for Hazard Evaluation Procedures,* American Institute of Chemical Engineers, AIChe, New York, USA, 1985.

26 Hale, A. R., and Glendon, A. I., *Individual behaviour in the Control of Danger*, Elsiever, Amsterdam, The Netherlands, 1987.
27 Hellstrand, T., *Drilling Shallow gas in the Norwegian Sector*, IADC European Well Control Conference, Esbjerg, Denmark, November 1990.
28 Hendrick, K., and Benner, L., *Investigating Accidents with STEP*. Marcel Dekker, Inc., New York, USA, 1987.
29 Holand, P., *Offshore Blowouts, Causes and Trends,* Ph.D. Thesis, Norwegian Institute of Technology, Trondheim, Norway, 1996.
30 Holand, P., *Offshore Blowouts, Data for Risk Assessment*, ASME paper No.OMAE-95-1333, Copenhagen, Denmark, June 1995.
31 Holand, P., *Reliability of Surface Blowout Preventers (BOPs),* SINTEF Report STF75 F91037, Trondheim, Norway, 1992.
32 Holand, P., "Subsea Blowout Preventer Systems, Reliability and Testing", *SPE Drilling Engineering, December 1991.*
33 Holand, P., *Subsea BOP Systems, Reliability and Testing Phase V, revision 1* (this revision 1 is based on a report with the same title published in 1990), SINTEF report STF 75 A89054, Trondheim, Norway, 1995.
34 Holand, P., "Well barrier analysis, blowout risk, reliability of subsea blowout preventers", Lecture at the EEU course, Safety and reliability of subsea production systems, Norwegian Institute of Technology, Trondheim, Norway 1993.
35 Hughes, V. M. P., Podio, A. L., Sepehrnoori, K, "A Computer-Assisted Analysis of Trends Among Gulf Coast Blowouts", *In Situ Journal*, 14(2), 201 - 228, USA, 1990.
36 *Hydril Compact Blowout Preventers, Lightweight Field Serviceable Systems, Brochure*, Hydril Company, Houston, TX, USA.
37 Høyland, A, *Sannsynlighetsregning og statistisk metodelære, del II metodelære,* Tapir, Trondheim, Norway, 1983 (in Norwegian).
38 Høyland, A., and Rausand, M., *System Reliability Theory; Models and Statistical Methods,* John Wileys Sons, New York, USA, 1994.
39 Johnson, W. G., *MORT Safety Assurance System*, Marcel Dekker, New York, USA, 1980.
40 Jones R. B., *Risk-Based Management, A reliability centred approach,* Gulf Publishing Company, Houston, TX, USA, 1995.

41 Kjellén, U., and Hovden, J., "Reducing risks by deviation control - a retrospection into a research strategy", *Safety Science*, 16:417-438, 1993.
42 Kjellén, U., Tinmannsvik, R. K., Ulleberg, T., Olsen, P. E., and Saxvik, B., *SMORT - Sikkerhetsanalyse av industriell organisasjon, offshore versjon*, Yrkeslitteratur, Oslo, Norway, 1987 (in Norwegian).
43 Lees, F. P., *Loss prevention in the process industries, Volume 1*, Butterworth Heineman Ltd, Oxford, United Kingdom, 1980.
44 Letter from H. G. Frizzell, Minerals Management Service, GoM OCS region, Reference MS5221, March 26, 1992, New Orleans, LA, USA.
45 Letter from Norwegian Petroleum Directorate, ref code OD 94/471-11/SA, September 8,1995, concerning Norwegian Offshore Accidents. NPD regarded the list of accidents enclosed to the letter unofficial and could not guarantee that all accidents were included.
46 Maclachlan, M., *An Introduction to Marine Drilling,* Dayton's, Oilfield Publications Limited, United Kingdom, 1987.
47 Martin W. H. *New MMS Drilling Regulations* SPE/IADC 18674, SPE/IADC Drilling Conference, New Orleans, LA, USA, Feb. 28-March 3, 1989.
48 Molnes E. M., *Reliability of Well Completion Equipment,* SINTEF Report STF75 F89030, Trondheim, Norway, 1989.
49 Molnes, E. M., *Reliability Analyses of Subsea and Well Systems*, Practical Examples Results and Limitations, Petrobras IV Technical Meeting - Reliability Engineering, Rio de Janeiro, Brazil, Aug. 7-10, 1995.
50 Moore, B., "Shallow gas hazards - the HSE Perspective", *Petroleum Review*, United Kingdom, September 1992
51 *NPD Annual Reports*, 1980 to 1994 editions, Norwegian Petroleum Directorate (NPD), Stavanger, Norway.
52 Offshore Accident and Incident Statistics Report 1994, *Offshore Technology Report* - OTO 95 953, Health and Safety Executive, Sheffield, United Kingdom, May 1995.
53 *Oil Spill Intelligence Report*, Vol. III, No. 4, Aug. 22,1980.
54 *Olie en gas in Nederland opsporing en winning 1993,* Ministry of Economic Affairs, Directorate General for Energy Information Directorate, The Hague, The Netherlands, April 1994.

55 Rike J. L. et al., *Completion and Workover Well Control Needs are Different, SPE 21563,* Calgary, Canada, June 10 -13, 1990.

56 Roche, J. R., Equipment for diverting shallow gas, Petro-Safe Pennwell Conference, USA, Oct. 3-5, 1989.

57 *ROCOF - PC Program for Analysis of Life Data for Repairable Systems - β version*, Developed by SINTEF Safety and Reliability, Trondheim, Norway, 1991.

58 Rosenberg, T., Nielsen, T. E., "Blowout Risk Modeling", OMAE Volume II, Safety and Reliability, Copenhagen, Denmark, June 1995.

59 Safety & Environmental Management Program (SEMP), Minerals Management Service, Internet address http://www.mms.gov, updated Oct. 10, 95.

60 Sewell, S., Director, Minerals Management Service, press release #30001, USA, January 1993.

61 *SINTEF Offshore Blowout Database,* April 1995 release, incl *Users' Manual*, Trondheim, Norway, 1995.

62 *Statistical Yearbook 1995*, Statistics Norway Oslo - Kongsvinger, Norway, 1995.

63 Surry, J., *Industrial Accident Research. A Human Engineering Appraisal,* Labour Safety Council, Ontario Ministry of Labour, Toronto, Canada, 1974.

64 Tallby R. J., *Barriers in Well Operations*, Statoil Report no R00290rjt/RJT, Stavanger, Norway, 1990.

65 Telephone conversation with Bartholomew, Henry G., deputy associate director for offshore operations and safety management, Minerals Management Service, Nov. 9, 1995.

66 Telephone conversation with Russ Hornzaa, Department of Trade and Industry, Production/injection data from 1991, 1992 and 1993.

67 The Bureau of Labour Statistics, Agency within the U.S. Department of Labour, accessed via Internet. Address http://stats.bls.gov/blshome.html, USA, 1995.

68 *The Energy report, Development of the Oil and Gas Resources of the United Kingdom, Volume 2, (Brown Book),* Department of Trade and Industry, London, United Kingdom, 1994.

69 *The North Sea Field Development Guide, 4th edition*, Oilfield Publications Limited (OPL) Ledbury, United Kingdom, June 1992.

70 *The Public Inquiry into the Piper Alpha Disaster (Cullen report),* Department of Energy, HMSO 1990 ISBN 0 10 113102 X, United Kingdom, 1990.

71 US GoM OCS wells drilled, Minerals Management Service, Internet address http://www.mms.gov/omm/gomr/homepg/pubinfo/freeasci/-well/freewell. html. The version of 27 June 1996 has been used.

72 *West Vanguard rapporten (The West Vanguard Report),* Norsk Offentlig Utredninger NOU report 1986:16, Norway (in Norwegian).

73 *WOAD, World Offshore Accident Databank, Statistical Report 1994*, DNV Technica, Norway, 1994.

74 *World Oil*, February issues (1980-1995), Gulf Publishing Company, Houston, TX, USA.

75 Østebø, R. et. al., *Shallow Gas and Leaky Reservoirs*, Norwegian Petroleum Society Conference, Stavanger, Norway, April 10-11, 1989.

INDEX

C

D

E

F

G

H

I

J

K

L

U

W

CPSIA information can be obtained at www.ICGtesting.com
Printed in the USA
LVOW04*1935160915

454136LV00007B/18/P

9 780884 155140